The Heavier *d*-Block Metals: Aspects of Inorganic and Coordination Chemistry

Catherine E. Housecroft

Series sponsor: **ZENECA**

ZENECA is a major international company active in four main areas of business: Pharmaceuticals, Agrochemicals and Seeds, Specialty Chemicals, and Biological Products.

ZENECA's skill and innovative ideas in organic chemistry and bioscience create products and services which improve the world's health, nutrition, environment, and quality of life.

ZENECA is committed to the support of education in chemistry and chemical engineering.

OXFORD
UNIVERSITY PRESS

OXFORD

UNIVERSITY PRESS

Great Clarendon Street, Oxford OX2 6DP
Oxford University Press is a department of the University of Oxford
and furthers the University's aim of excellence in research, scholarship,
and education by publishing worldwide in

Oxford New York

Athens Auckland Bangkok Bogotá Buenos Aires Calcutta
Cape Town Chennai Dar es Salaam Delhi Florence Hong Kong Istanbul
Karachi Kuala Lumpur Madrid Melbourne Mexico City Mumbai
Nairobi Paris São Paulo Singapore Taipei Tokyo Toronto Warsaw

and associated companies in Berlin Ibadan

Oxford is a registered trade mark of Oxford University Press

Published in the United States
by Oxford University Press Inc., New York

A catalogue record for this title is available from the British Library
Data available

Library of Congress Cataloging in Publication Data
Data available
ISBN 0–19–850103X (Pbk)

Typeset by the author

Printed in Great Britain
on acid-free paper by
The Bath Press, Bath

Series Editor's Foreword

Oxford Chemistry Primers are designed to give a concise introduction to all chemistry students by providing the material that would usually form an 8–10 lecture course. As well as providing up-to-date information, this series expresses the explanations and rationales that form the framework of current understanding of inorganic chemistry.

Catherine Housecroft has provided an essential guide to second and third row transition element chemistry, which incorporates a fine overview of the structures and reactivity of this important part of the Periodic Table.

Much of Catherine's own research has involved these elements, giving her an excellent background for writing this text. Her previous Primers have been noteworthy for the high quality of their presentation, and this book is a worthy addition to those.

Companion books covering the first row transition elements and the f-block elements will succeed this volume. Transition-element chemistry is universal in undergraduate chemistry courses, and this suite is perhaps the most approachable and up-to-date companion to these courses.

John Evans
Department of Chemistry,
University of Southampton

Preface

This short book is designed as an introduction to the inorganic and coordination chemistry of the second and third row *d*-block metals, and it is assumed that students using this book will already have a grounding in the chemistry of the first row *d*-block metals. It has not been easy to choose the scope of coverage for this *Primer*; within the remit of the title of the book lies a wealth of chemistry far in excess of the page limit. In deciding on the content, I have been mindful, firstly, that second and third row metals are often used to introduce students to special topics (e.g. high coordination numbers, metal–metal bonding, polyoxometallates) and, secondly, to avoid overlap with other *Primers*. I have supplemented the book with many suggestions for further reading, including papers in the current literature both from research and more topical journals, and I hope that these articles will stimulate students to find out more about the heavier metals.

I must thank Professor John Evans and those at OUP for their patience in waiting for me to complete this project. As always, I am extremely grateful to my husband, Professor Edwin Constable, for his role as an excellent critic. And, last but not least, thanks once again go to Philby and Isis for their peaceful (and not so peaceful) feline company; writing a book would not be the same without them.

Basel C. E. H.
September 1998

Contents

1 The metals: occurrence, physical properties and uses

1.1 The second and third row *d*-block metals

The aims of this book are to introduce representative inorganic and coordination chemistry of the heavier metals of the *d*-block. Figure 1.1 shows the positions of these metals within the periodic table. In group 3, lanthanum (La) is generally considered with the *f*-block elements. Notice in Fig. 1.1 that whereas the atomic numbers of the elements in the second row of the *d*-block follow sequentially, there is a discontinuity in the atomic numbers between La and Hf in the third row. In Chapter 2 we discuss the effects of introducing the lanthanoid elements before hafnium. The ground state electronic configurations of the metals listed in Table 1.1 also illustrate the introduction of the lanthanoid series.

The metals ruthenium, osmium, rhodium, iridium, palladium and platinum are referred to collectively as the 'platinum-group metals' and occur together naturally.

The 1993 recommendations of the IUPAC name the elements from La to Lu as the *lanthanoids* rather than the lanthanides.

The platinum-group metals

Ru	Rh	Pd
Os	Ir	Pt

1-2	3	4	5	6	7	8	9	10	11	12	13-18
s-block	Sc 21	Ti 22	V 23	Cr 24	Mn 25	Fe 26	Co 27	Ni 28	Cu 29	Zn 30	*p*-block
	Y 39	Zr 40	Nb 41	Mo 42	Tc 43	Ru 44	Rh 45	Pd 46	Ag 47	Cd 48	
	La 57	Hf 72	Ta 73	W 74	Re 75	Os 76	Ir 77	Pt 78	Au 79	Hg 80	

Fig. 1.1. The heavier *d*-block elements are highlighted. Atomic numbers are given.

Table 1.1 Ground state electronic configurations of the second and third row *d*-block elements

Element	Symbol	Ground state electronic configuration
Yttrium	Y	$[Kr]5s^2 4d^1$
Zirconium	Zr	$[Kr]5s^2 4d^2$
Niobium	Nb	$[Kr]5s^1 4d^4$
Molybdenum	Mo	$[Kr]5s^1 4d^5$
Technetium	Tc	$[Kr]5s^2 4d^5$
Ruthenium	Ru	$[Kr]5s^1 4d^7$
Rhodium	Rh	$[Kr]5s^1 4d^8$
Palladium	Pd	$[Kr]5s^0 4d^{10}$
Silver	Ag	$[Kr]5s^1 4d^{10}$
Cadmium	Cd	$[Kr]5s^2 4d^{10}$
Lanthanum	La	$[Xe]6s^2 5d^1$
Hafnium	Hf	$[Xe]4f^{14} 6s^2 5d^2$
Tantalum	Ta	$[Xe]4f^{14} 6s^2 5d^3$
Tungsten	W	$[Xe]4f^{14} 6s^2 5d^4$
Rhenium	Re	$[Xe]4f^{14} 6s^2 5d^5$
Osmium	Os	$[Xe]4f^{14} 6s^2 5d^6$
Iridium	Ir	$[Xe]4f^{14} 6s^2 5d^7$
Platinum	Pt	$[Xe]4f^{14} 6s^1 5d^9$
Gold	Au	$[Xe]4f^{14} 6s^1 5d^{10}$
Mercury	Hg	$[Xe]4f^{14} 6s^2 5d^{10}$

1.2 Occurrences, physical properties and uses

In this section, we survey some of the properties and uses of the second and third row metals. A list of articles detailing further aspects of uses or potential applications is given in 'metals in action' on page. 88.

Metals in Section 1.2 are arranged in order of increasing atomic number; lanthanum is omitted.

1.2.1 Yttrium

Yttrium occurs in the form of the ores *monazite* (a mixed metal phosphate containing about 3% yttrium) and *bastnäsite* (which contains ≈0.2% yttrium); commercial production involve reactions 1.1 and 1.2.

$$2YF_3 + 3Ca \rightarrow 2Y + 3CaF_2 \tag{1.1}$$

$$YCl_3 + 3K \rightarrow Y + 3KCl \tag{1.2}$$

Where a metal possesses an NMR active nucleus with $I = \frac{1}{2}$, the % abundance of the active nucleus has been given. In not all cases are these nuclei necessarily observed *directly* as mentioned in the text. Other nuclei are also NMR active, e.g. ^{181}Ta, $I = \frac{7}{2}$; ^{193}Ir, $I = \frac{3}{2}$

Yttrium is a silver-coloured and brittle metal; it is fairly stable in air in the bulk state but metal turnings ignite if heated above 670 K. It is used as Y_2O_3 in the manufacture of phosphors in television tubes, in the formation of yttrium garnets for microwave filters, and in synthetic gemstones (yttrium aluminium garnets, YAG). The only naturally occurring isotope of yttrium ^{89}Y is NMR active and exhibits a very wide shift range (>1000 ppm) making the use of ^{89}Y NMR spectroscopy valuable in characterizing yttrium-containing compounds.

NMR spectroscopic data:

^{89}Y 100% $I = {}^1/_2$

1.2.2 Zirconium

After iron, titanium and manganese, zirconium is the next most abundant *d*-block metal in the Earth's crust; lunar rock samples collected by the Apollo missions have been shown to contain relatively high amounts of zirconium. Zirconium occurs naturally with hafnium and their separation is difficult. The major ores are *zircon* ($ZrSiO_4$) and *baddeleyite* (ZrO_2), and extraction processes involve the reduction of the oxide by calcium, reduction of K_2ZrF_6 produced from zircon by treatment with K_2SiF_6, or reduction of $ZrCl_4$ by magnesium. Zirconium is a lustrous and ductile, silvery metal which has excellent resistance to corrosion. This property is the key to many of its uses and a major application is in the cladding of fuel rods in water-cooled nuclear reactors. Zirconium is also a component in a range of catalytic systems.

1.2.3 Niobium

Niobium occurs naturally in the ore *pyrochlore* $NaCaNb_2O_6F$ and in *columbite* (also called *niobite*) which has varying compositions of the general formula $(Fe,Mn)(Nb,Ta)_2O_6$. In its manufacture, niobium is usually separated from tantalum by a liquid extraction process, extracting into $Me_3CCH_2CH_2OMe$ from HF solutions. Niobium is a shiny, white, soft and ductile metal; it oxidizes in air above 470 K. It is used in the manufacture of certain grades of stainless steel and has found application in frameworks designed for the Gemini space program; it is also used in superconducting magnets.

Niobite and *tantalite* (Section 1.2.12) appear to have the same formula, but are distinguished by the variable composition — niobite is niobium-rich, while tantalite is tantalum-rich.

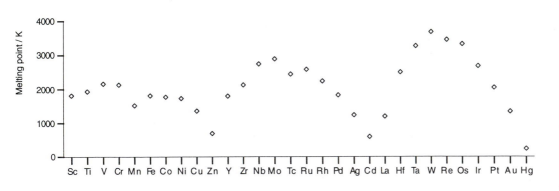

Fig. 1.2. Trends in melting points of the *d*-block elements.

1.2.4 Molybdenum

The main ore from which molybdenum is extracted is *molybdenite* MoS_2; roasting MoS_2 in air produces MoO_3 which is then reduced with H_2. Molybdenum is a very hard metal with a high melting point (Fig. 1.2); it is used extensively in alloys (e.g. steels) where it provides increased strength and hardness.

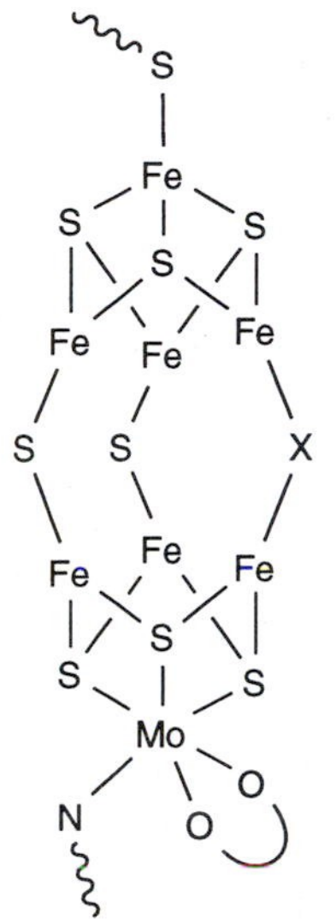

Molybdenum has an essential role in biological systems where it occurs in the active sites of *molybdoenzymes*. These fall into two general classes: the nitrogenases which involve the iron-molybdenum cofactor FeMoco, and the molybdopterin-containing enzymes. Molybdenum-containing nitrogenases catalyse the reduction of N_2 to NH_3 (nitrogen fixation), for example, by leguminous plants as part of the nitrogen cycle. Nitrate reductases reduce nitrate to nitrite when plants assimilate nitrogen (e.g. from fertilizers). Since 1992, X-ray crystallographic data have been available that provide detailed structural data concerning the iron-molybdenum cofactor. Figure 1.3 shows the Fe/Mo/S-cluster that is tethered to a protein by an *S*-coordinated cystenyl residue, and an *N*-bound histidine unit; the chelating *O,O'*-unit in Fig. 1.3 represents an isocitrate unit. The identity of atom X in the figure is not known with certainty, but it may be a third sulfido-bridge.

Fig. 1.3. Fe/Mo/S-cluster core of the iron-molybdenum cofactor.

When an element is found *native*, it is found 'in the elemental state'.

1.2.5 Technetium

Technetium is an artificial element and the isotope most commonly encountered in the laboratory is ^{99}Tc, a β-emitter with a half-life of 2.1×10^5 years. ^{99}Tc is a fission product of ^{235}U and is produced in the form of the pertechnetate(VII) ion $[TcO_4]^-$. The metastable ^{99m}Tc isotope (a γ-emitter with a half life of 6.02 hours) is the most important imaging agent in nuclear medicine. Technetium is a shiny silver-coloured metal which becomes superconducting below 11 K.

1.2.6 Ruthenium

Native ruthenium occurs with other platinum metals and also in the ores *pentlandite* and *pyroxinite*. It is a hard, white metal and oxidizes in air only above 1070 K; it resists attack by acids but reacts with halogens and alkalis. Commercially, it is most often purchased as 'hydrated ruthenium trichloride' $RuCl_3 \cdot nH_2O$. Uses of ruthenium include alloying with platinum and palladium to increase their hardness, for example in the production of electrical components, and it has widespread applications in a range of catalytic systems.

1.2.7 Rhodium

Rhodium occurs native with other platinum metals and its low abundance (only 1×10^{-4} ppm of the Earth's crust) renders it expensive on the commercial market. It is a hard, silver-white metal, stable with respect to oxidation in air up to ≈ 875 K, and resistant to attack by acids. It is used as an alloying component with platinum and palladium and these hardened alloys have wide applications, e.g. in thermocouples, electrodes and laboratory crucibles. Catalytic applications include those in motor-vehicle catalytic converters.

The only stable isotope of rhodium, ^{103}Rh, is NMR active. This property is invaluable in aiding the characterization of rhodium-containing compounds, *not* necessarily by direct observation of ^{103}Rh NMR spectra but by the observation of coupling to other nuclei such as ^{1}H, ^{13}C or ^{31}P.

NMR spectroscopic data:

^{103}Rh 100% $I = {}^1/_2$

1.2.8 Palladium

Palladium occurs with other platinum metals; it is a grey-white, malleable and ductile metal but its strength can be increased by cold-working. Palladium does not oxidize in air, but it reacts slowly with sulfuric and nitric acids. Major applications of palladium are in the electronics industry where it is used in printed circuits and multilayer ceramic capacitors. It has widespread use as a catalyst in the laboratory and in industry, with its role in catalytic converters being of prime importance for the environment. An unusual property of palladium is its ability to absorb large amounts of H_2 and this leads to its application in the purification of dihydrogen on an industrial scale.

Cold fusion : the fusion of hydrogen nuclei to give helium is the source of the Sun's energy; however, the conditions needed to initiate fusion are harsh. In 1989, Pons and Fleischmann reported the remarkable result that they had achieved fusion of deuterium nuclei under ambient conditions through the electrolysis of D_2O using Pd electrodes. The formation of D_2 in this experiment is followed by absorption of the deuterium by the palladium, and it was reported that large amounts of energy were unexpectedly produced by the system and this was interpreted as co-called 'cold fusion' of deuterium nuclei. Attempts to reproduce the results have not met with success and, to date, the phenomenon of cold fusion remains in dispute.

1.2.9 Silver

The symbol Ag for silver is derived from the Latin name for the element *argentum*. Silver occurs as a native metal and as the ore *argenite* Ag_2S; it is also recovered during the manufacture of other metals such as copper. Pure silver is a lustrous silver-white metal, and has the highest electrical conductivity of any metals. Silver tarnishes in air, forming the black sulfide, and the alloy *sterling silver*, which contains 92.5% Ag, tends to be used in jewellery and cutlery manufacture. Silver reacts with oxidizing acids (e.g. with hot concentrated H_2SO_4) but resists attack by alkalis. It has extensive uses in the photographic industry, and also finds applications in the manufacture of high capacity batteries, soldering-alloys, electrical equipment and printed circuits. Each of the isotopes ^{107}Ag and ^{109}Ag is NMR active, but their use in compound characterization usually comes in the detection of coupling to other nuclei, e.g. ^{31}P.

NMR spectroscopic data:

^{107}Ag 51.8% $I = {}^1/_2$

^{109}Ag 48.2% $I = {}^1/_2$

1.2.10 Cadmium

The only major ore of cadmium is *greenockite* (CdS) but more generally, cadmium occurs in association with zinc-bearing ores and is usually produced as a by-product of the manufacture of zinc. Cadmium is a soft, ductile, blue-white metal; Fig. 1.2 illustrates its low melting point, consistent with other members of group 12, and this property leads to its use as an alloying agent in the manufacture of low melting alloys. It oxidizes in air, and reacts with dilute acids. Cadmium is a toxic element, as are solutions of its salts, and is recognized as a possible carcinogen. Cadmium selenide (CdSe) and telluride (CdTe) are semiconductors and have widespread application in the electronics industry.

Two of the eight naturally occurring isotopes of cadmium are NMR active (^{111}Cd and ^{113}Cd) and ^{113}Cd in particular is readily observed. Perhaps more commonly, use is made of the observation of satellite peaks in, for example, ^{1}H NMR spectra.

NMR spectroscopic data:

^{111}Cd 12.8% $I = {}^1/_2$

^{113}Cd 12.3% $I = {}^1/_2$

1.2.11 Hafnium

Hafnium is a ductile, silver-coloured metal. It occurs naturally in association with zirconium — the ore *zircon*, formulated as $ZrSiO_4$, contains <2% Hf, and *alvite*, $MSiO_4 \cdot xH_2O$ (M = Hf, Zr) contains similar ratios of Zr:Hf. An extraction process involves the reduction of $HfCl_4$ (produced from zircon) by magnesium; see also Section 1.2.2. Whereas zirconium has a *low* neutron capture cross-section and is used in the cladding of fuel rods in nuclear reactors, hafnium has a *high* neutron capture cross-section (i.e. it absorbs neutrons efficiently) and its major application is in nuclear reactor control rods.

1.2.12 Tantalum

The principal ore of tantalum is *tantalite* $(Fe,Mn)(Nb,Ta)_2O_6$. The metal is hard, ductile, has a high melting point (Fig. 1.2) and is extremely resistant to corrosion by air or water; these properties make it valuable as an alloying agent. Some of its major applications are in capacitors and in tantalum carbide cutting tools, and its inertness within the body has led to its use in surgical appliances including prostheses.

1.2.13 Tungsten

In the German language, tungsten is called *wolfram*, and it is from this name that the symbol W derives. The metal occurs naturally in the form of the ores *wolframite* $(Fe,Mn)WO_4$ and *scheelite* $(CaWO_4)$, and is extracted on an industrial scale by the formation of WO_3 which is subsequently reduced by H_2. Tungsten is a silver-white metal and has the highest melting point of the *d*-block metals (Fig. 1.2); it has important applications in electrical filaments including those in household bulbs, and in steel alloys — tungsten carbides are used extensively for cutting tools.

Of the five naturally occurring isotopes, ^{183}W is NMR active and the appearance of satellite peaks (Fig. 1.4) in, for example, ^1H and ^{31}P NMR spectra, is a useful observable in the characterization of tungsten-containing compounds.

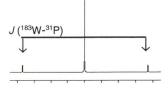

$J\,(^{183}W\text{-}^{31}P)$

Fig. 1.4. In a ^{31}P NMR spectrum of a compound containing a P–W bond, satellites due to spin-spin coupling between ^{31}P and ^{183}W are observed.

NMR spectroscopic data:

^{183}W 14.3% $I = {}^1/_2$

1.2.14 Rhenium

Rhenium is a rare metal, occurring to an extent of only 0.0007 ppm in the Earth's crust. There is no ore of significant importance, and industrially, rhenium is extracted from flue dusts collected after molybdenite has been roasted (see Section 1.2.4). The dust is oxidized with hypochlorite to produce $[ReO_4]^-$, which is then reduced to the metal. Rhenium is a dense, very high melting (Fig. 1.2), silver-grey metal. It is alloyed with molybdenum and tungsten to provide commercially useful alloys; rhenium-tungsten alloys are used for thermocouples. Catalytic applications of rhenium are important, as are its uses in filaments for photographic flash-equipment and mass spectrometers.

1.2.15 Osmium

Osmium is a rare metal (abundance 0.005 ppm of the Earth's crust) and occurs with other platinum-group metals (see Section 1.2.16). It is readily

oxidized to the volatile, colourless, highly toxic oxide OsO_4 and this accounts for the characteristic odour of the metal — the name *osmium* is derived from the Greek word *osme* meaning 'smell'. The metal is lustrous, blue-white, and extremely hard, dense and high-melting (Fig. 1.2). Commercial uses of osmium are limited but it has some application as an alloying agent.

1.2.16 Iridium

Like osmium, iridium is rare and expensive, and a major source is *osmiridium*, a native alloy comprising 15-40% osmium and 50-80% iridium. The densities of these two metals are virtually identical and they are the densest elements known. Iridium is lustrous, hard and brittle, and is difficult to work in its pure state; it is extremely resistant to corrosion, and is not attacked by air, water or acids. The metal has limited uses; it is alloyed with platinum to impart increased hardness to the latter, and is used in the form of an osmium-iridium alloy in pen-nibs.

There are two natural isotopes of iridium, ^{191}Ir and ^{193}Ir. Although both are NMR active, the large quadrupole moments preclude significant application in NMR spectroscopy. The ^{193}Ir nucleus is however suitable for Mössbauer spectroscopy.

1.2.17 Platinum

Native platinum occurs associated with other platinum-group metals, and *sperrylite* ($PtAs_2$) is an important platinum-bearing ore. Platinum is a silver-coloured metal, malleable and ductile, and is widely used for the manufacture of wires, electrodes, jewellery and thermocouples. It is inert in air, withstanding oxidation even at high temperatures. It has an important role as a catalyst, both for chemical transformations in industry and in motor vehicle catalytic converters. Clinical uses of platinum-containing compounds are significant; cisplatin **1.1** has been applied as an anti-tumour drug since the 1960s and carboplatin **1.2** is similarly used with the advantage that it appears to have fewer side effects than cisplatin.

Of the six natural isotopes of platinum, ^{195}Pt is NMR active. The observation of satellite peaks in, for example, ^{1}H, ^{31}P or ^{19}F NMR spectra provides valuable structural information about platinum-containing compounds. Additionally, coupling constants for directly attached nuclei are large and long range coupling is readily observed.

1.2.18 Gold

Gold occurs to an extent of 0.004 ppm in the Earth's crust both in the elemental state and in telluride ores, and is extracted via the formation of the linear anion $[Au(CN)_2]^-$ **1.3**. Gold is yellow in colour, and is soft, malleable and ductile, and resists atmospheric oxidation and attack by almost all acids; it is, however, dissolved by *aqua regia*, a 1:3 mixture of HNO_3 and HCl. Its combination with other metals to give alloys of increased strength is often required, and use of the term *carat* indicates the gold content (24 carat = pure gold). Uses of gold are varied, and include jewellery, coinage and the

(1.1)

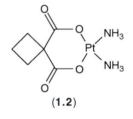

(1.2)

NMR spectroscopic data:

^{195}Pt 33.8% $I = ^1/_2$

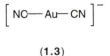

(1.3)

electronics industry; certain compounds including disodium thiomalate $Na_2[O_2CCH_2CH(SAu)CO_2]$ are used as anti-rheumatic drugs. Suspensions of *colloidal gold* (formed by reduction of gold chloride) may be red, blue or purple and are used in electron microscope imaging and to stain microscope slides as well as colouring agents.

1.2.19   Mercury

Cinnabar (HgS) is the main ore from which mercury is obtained and the metal is extracted by roasting the ore in air. Figure 1.2 showed that mercury possesses the lowest melting point of the *d*-block metals and the fact that mercury is a liquid at room temperature is well known — the symbol Hg derives from the Latin word *hydrargyrum* meaning 'liquid silver'. Mercury is used in thermometers, barometers, laboratory diffusion pumps and electrical equipment including switching devices. It is however a cumulative poison and is readily absorbed by the body. Intermetallic *amalgams* are formed between mercury and a range of other metals; Na/Hg amalgam is a commonly encountered reducing agent, and Cd/Hg amalgam is the cathode in the Weston standard cell:

$$Cd\,(Hg)\,\big|\,CdSO_4, H_2O \vdots Hg_2SO_4\,\big|\,Hg$$

2 Periodicity

2.1 Moving from the first to second and third rows of the *d*-block

In textbooks, the chemistry of the first row *d*-block metals is often discussed separately from that of the metals in the second and third rows. Whereas the chemical properties of the heavier metals in a given triad are comparable, they show significant differences from those of the first member of the group. For example, trends in oxidation states and geometrical preferences partition each triad into two categories: (i) the first member, and (ii) the second and third members. Discussions which attempt to encompass a whole triad usually run into difficulties, although, as one might expect from a comparison of the ground state electronic configurations, there are *some* chemical similarities between the metals in a given group. For example, the maximum oxidation state attained by each of Cr, Mo and W is +6; however, as we shall see later, the stability of this oxidation state is greater for Mo and W than for Cr. Similar patterns of behaviour are observed for the other triads.

In the rest of this chapter, we consider properties of the *d*-block metals that emphasize the distinction between the first and the two heavier members of each triad, and also consider trends across the second and third periods of the *d*-block.

2.2 Metallic and ionic radii

Figure 2.1 shows the trends in metallic radii, r_m, for the *d*-block metals. The value of r_m is equal to half of the internuclear separation in the crystal lattice; the assumption is made that the metal atoms are spheres that touch each other. Values of r_m in Fig. 2.1 refer to twelve-coordinate metal atoms; this coordination number is relevant to a close-packed array of metal atoms. Many metals are *polymorphic* and it may be necessary to quench a high-temperature form to room temperature in order to determine interatomic distances for the high-temperature polymorph. However, if only the metallic radius for an eight-coordinate (body-centred cubic) metal atom has been experimentally determined, it is possible to use the relationship in Eqn 2.1 to estimate the metallic radius appropriate for a twelve-coordinate metal atom.

If a substance is polymorphic, it exists in more than one crystal form. For example, under a pressure of 1 bar, iron possesses a body-centred cubic (bcc) lattice up to 1183 K, a cubic close-packed lattice over the temperature range 1183 to 1663 K, and a bcc lattice above 1663 K.

$$\text{For a given metal}: \quad \frac{r_{\text{8-coordination}}}{r_{\text{12-coordination}}} \approx 0.97 \qquad (2.1)$$

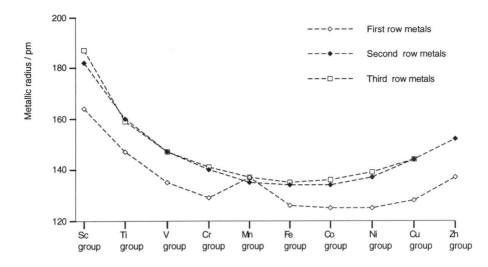

Fig. 2.1. Trends in metallic radii of the *d*-block metals; mercury is excluded from the data.

Figure 2.1 shows that, with the exception of manganese, each first row metal is smaller than its heavier congeners. However, there is little difference in size between the second and third members of a given triad; that is, an increase in r_m on going from a second to third row metal which would be expected on the grounds of an increase in the principal quantum number and an increased number of electrons, is not observed. The explanation for this lies in the so-called *lanthanoid contraction* and this influences trends in metallic, ionic and covalent radii. As we have already seen (page 1), the series of fourteen lanthanoid elements are positioned between lanthanum ($Z = 57$) and hafnium ($Z = 72$); lanthanoids are characterized by possessing a $4f^n$ ground state electronic configuration, and as the $4f$ level is filled, the metals show a steady decrease in size. The contraction arises because of the poor shielding effects of the $4f$ electrons on one another, and the consequently greater effective nuclear charge experienced by the valence electrons.

2.3 Coordination numbers

The larger size of the heavier elements of the *d*-block means that, compared to their first row counterparts, higher coordination numbers are observed. Examples include $[ZrF_7]^{3-}$ and $[Mo(CN)_8]^{2-}$ which possess monocapped trigonal prismatic and eight-coordinate structures, respectively. The structural diagrams in Fig. 2.2, include views in which the ligand polyhedra are delineated. Further examples of high coordination numbers are given later in the book.

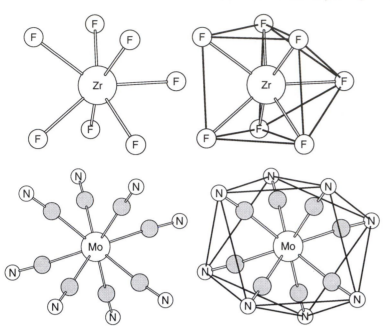

Fig. 2.2. The structures of $[ZrF_7]^{3-}$ and $[Mo(CN)_8]^{4-}$ determined by X-ray diffraction studies for the guanidinium and tetra-*n*-butylammonium salts respectively.

Although coordination numbers such as seven, eight and nine are observed, lower values are important as in the first row, with four-, five- and six-coordinate metal centres being common.

2.4 Oxidation states

In *organometallic* compounds containing π-acceptor ligands (e.g. CO), *low* oxidation states are stabilized on descending a triad; for example, more osmium and ruthenium cluster species of general type $[M_x(CO)_y]$ are known than corresponding iron species. For non-organometallic compounds, the stability of *higher* oxidation states *increases* down the group, and, in general, lower oxidation states are less stable for the heavier than the lighter elements. This is well illustrated by consideration of group 6. For chromium, +3 is a common oxidation state, and Cr(VII) (e.g. as the dichromate ion, $[Cr_2O_7]^{2-}$ in acidic solution) is a powerful oxidizing agent (Eqn 2.2). In contrast, Mo(VI) and W(VI) (Eqn 2.3) are relatively stable with respect to reduction, while Mo(II), Mo(III), W(II) and W(III) are uncommon in mononuclear compounds.

$$[Cr_2O_7]^{2-}(aq) + 14H^+(aq) + 6e^- \rightleftharpoons 2Cr^{3+}(aq) + 7H_2O(l)$$

$$E^o = +1.33 \text{ V} \qquad (2.2)$$

$$WO_3(s) + 6H^+(aq) + 6e^- \rightleftharpoons W(s) + 3H_2O(l) \qquad\qquad E^o = -0.09 \text{ V} \qquad (2.3)$$

Redox reactions in aqueous solutions are discussed in Chapter 3.

2.5 Standard enthalpies of atomization and metal–metal bonding

A general observation in comparing the chemistry of first row d-block metals with their second and third row congeners is that compounds involving metal–metal bonding are more plentiful for the heavier metals. This phenomenon can be understood in terms of the effectiveness of metal–metal bonding interactions. Unfortunately, experimentally measured, *reliable* M–M bond dissociation enthalpy data are scarce. However, the standard enthalpy of atomization of a bulk metal provides information about the strength of metal–metal bonding in the element, although we must remember that these values refer to the zero valent state. The variation in $\Delta_{atom}H^o(298 \text{ K})$ for the d-block metals is shown in Fig. 2.3 and from these data we can gain some idea about the variation in metal–metal bond dissociation enthalpies.

Note however, that there are also structural changes to be taken into account; at 298 K, all the metals have close-packed lattices (hcp or ccp) *except for* metals in groups 5 and 6 (bcc), manganese (which has an unusual cubic lattice), iron (bcc), and mercury, which, of course is a liquid.

Below its melting point, mercury possesses a distorted, simple cubic structure.

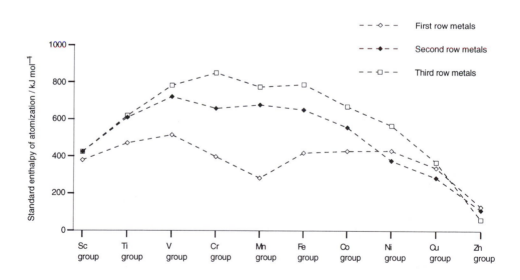

Fig. 2.3.   Trends in values of $\Delta_{atom}H^o$ (298 K) for the d-block metals.

Two general patterns emerge from Fig. 2.3.

- Values of $\Delta_{atom}H^\circ$(298 K) for the third row metals are greater than those for the second row which, in turn, are greater that those of the first row metals. This corresponds to greater orbital overlap between AOs of higher principal quantum numbers $5d$-$5d$ > $4d$-$4d$ > $3d$-$3d$.

- For the second and third rows, the highest values of $\Delta_{atom}H^\circ$(298 K) occur for the metals in the middle groups of the d-block. Significantly, it is for these metals that the majority of high-nuclearity, low oxidation state metal carbonyl clusters are observed.

For a discussion of metal–metal bonded carbonyl clusters, see: C.E. Housecroft (1996) *Metal–Metal Bonded Carbonyl Dimers and Clusters*, OUP, Oxford.

The importance of metal-metal bonding in dimetal complexes such as $[Mo_2(O_2CMe)_4]$ (Fig. 2.4) and in cluster species such as $[Mo_6Cl_8]^{4+}$ will be discussed in Chapters 6 and 7. Not all high nuclearity species, however, owe their existence to *direct* metal–metal bonded interactions, and the role of bridging ligands (such as halo or oxo bridges) is especially important, for example in $[Mo_6O_{19}]^{2-}$ (Fig. 2.4); species of this type are also discussed in Chapter 7.

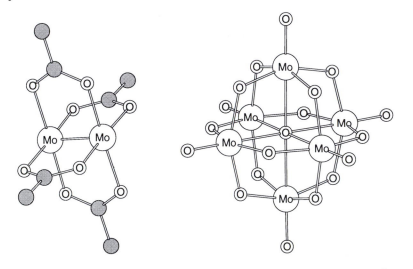

Fig. 2.4. The structures of $[Mo_2(O_2CMe)_4]$ (H atoms omitted) and of $[Mo_6O_{19}]^{2-}$, determined by X-ray diffraction methods, the latter for the tetrabutylammonium salt.

2.6 Summary

In going from the first row to the heavier metals in the d-block, significant differences are: increased atom and ion sizes, availability of higher coordination numbers, stability of higher oxidation states, and greater tendency towards metal–metal bonding. Additionally, there are differences in electronic spectroscopic and magnetic properties, as we discuss in Chapter 5.

3 Aqueous solution species

3.1 Introduction: an overview of first row metal behaviour

In this chapter, we consider two particular aspects of the chemistry of the second and third row *d*-block metals in aqueous solution:

- species present in solution;
- redox behaviour.

Throughout this chapter, the reader should keep in mind that the presence of coordinating ligands (including simple ions such as chloride) will influence species present in solution; (coordination complexes, see Chapters 4 and 5).

By way of introduction, the reader should review the species present in aqueous solutions containing metals ions from the first row of the *d*-block, and some important points are summarized below.

For the M^{2+} ions, octahedral hexaaqua ions such as $[Co(H_2O)_6]^{2+}$ (**3.1**) are formed for each metal, but oxidation to M^{3+} readily occurs for Ti, V, Cr and Fe. In each of the $[M(H_2O)_6]^{3+}$ ions so-formed, the polarizing power of the M^{3+} centre results in ready loss of a proton making these cations acidic in aqueous solution (Eqn 3.1, $pK_a = 2.0$). The hexaaqua ion can be stabilized in solution by addition of H^+ (i.e. application of Le Chatelier's principle).

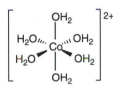

(**3.1**)

$$[Fe(H_2O)_6]^{3+}(aq) + H_2O(l) \rightleftharpoons [Fe(H_2O)_5(OH)]^{2+}(aq) + [H_3O]^+(aq) \qquad (3.1)$$

The increase in electron withdrawing power of the metal centre associated with an increase in oxidation state is further reflected in the fact that the hexaaqua ion of vanadium(IV) does not exist in solution, loss of two protons instead leading to the formation of $[V(H_2O)_4O]^{2+}$ (**3.2**). This is commonly written as $[VO]^{2+}$ with the presence of the water ligands being ignored, i.e. a parallel shorthand to writing Co^{2+} for cation (**3.1**).

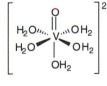

(**3.2**)

Loss of protons through polarization of O–H bonds in coordinated water leads, in some cases, to the formation of hydroxo ligands and the concomitant generation of dinuclear species. Such an example is $[(H_2O)_4Cr(\mu\text{-OH})_2Cr(H_2O)_4]^{4+}$ (**3.3**). Note the use of the Greek letter μ ('mu') to indicate the bridging nature of the hydroxo ligand. Without an added subscript, μ indicates that the ligand bridges between *two* metal centres; the symbol μ_3 indicates a triply bridging mode, as in Figure 3.7.

Although we stated above that water ligands are often ignored when writing formulae, it is important to remember their presence. For example, the reaction of aqueous nickel(II) ions with ammonia to give a complex ion is a *ligand displacement (substitution)* reaction. Another example comes in values of standard reduction potentials. Equation 3.2 refers to the reduction of the *hydrated* iron(III) ion. On the other hand, in Eqn 3.3, the presence of the

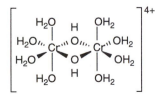

(**3.3**)

cyano ligands is implicitly stated. The effect that the change in ligands has on the relative ease of iron(III) reduction is dramatic.

$$Fe^{3+}(aq) + e^- \rightleftharpoons Fe^{2+}(aq) \qquad\qquad E^o = +0.77 \text{ V} \qquad (3.2)$$

$$[Fe(CN)_6]^{3-}(aq) + e^- \rightleftharpoons [Fe(CN)_6]^{4-}(aq) \qquad\qquad E^o = +0.36 \text{ V} \qquad (3.3)$$

All E^o values quoted in this book are with respect to the standard hydrogen electrode.

3.2 Potential diagrams

Although standard reduction potentials are readily available in tabulated form, the use of potential diagrams is a valuable method of displaying the redox behaviour of different species containing a particular element.

Figure 3.1 shows a potential diagram for tungsten species in acidic solution; pH 0 corresponds to $[H^+] = 1$ mol dm^{-3}, i.e. standard conditions. Each value of E^o applies to the *reduction* step indicated. For example, the value of -0.09 V corresponds to the standard reduction potential for reaction 3.4.

Exercise: assign an oxidation state to each metal in each species in Figs. 3.1, 3.2 and 3.3.

$$WO_3 + 6H^+ + 6e^- \rightleftharpoons W + 3H_2O \qquad\qquad\qquad (3.4)$$

Fig. 3.1. Potential diagram for tungsten in acidic solution (pH 0); E^o values in V.

The species shown in Fig. 3.1 for tungsten (group 6) show a marked contrast to those in the potential diagram for the first metal in the triad, chromium. Whereas lower oxidation states, e.g. Cr^{3+}(aq), are well established in aqueous solution for chromium (Fig. 3.2), this is not the case for tungsten. Notice, too, how the values of E^o in Fig. 3.2 indicate the high oxidizing power of Cr(VI) in acidic (pH 0) solution.

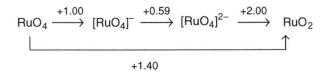

Fig. 3.2. Potential diagram for chromium in acidic solution (pH 0); E^o values in V.

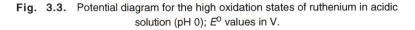

The state of protonation of oxoanions is pH dependent.

Fig. 3.3. Potential diagram for the high oxidation states of ruthenium in acidic solution (pH 0); E^o values in V.

The change in free energy under standard conditions, ΔG^0, is related to the standard electrode potential by the equation:

$$\Delta G^0 = -zE^0F$$

where:

ΔG^0 is in J mol^{-1}
E^0 is in V
z = electrons transferred
F = Faraday constant = 96 485 C

The individual steps in potential diagrams may involve one or more electrons, and this is important to remember when comparing reduction potentials for parallel groups of steps. This is illustrated in the detailed sequence from RuO_4 to Ru in Fig. 3.3; for example, the reduction of RuO_4 (Ru(VIII)) to RuO_2 (Ru(IV)) is described in terms of a single step with E^0 = +1.40 V, or in terms of a series of three steps. In order to equate the three separate reduction potentials to the overall value of +1.40 V, we must take into account the number of electrons transferred in each step; whereas values of E^0 *cannot* be summed directly, values of ΔG^0 *can* be summed. For the following four steps:

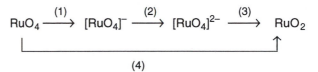

we can write:

$$\Delta G^0(4) = \Delta G^0(1) + \Delta G^0(2) + \Delta G^0(3)$$

$$-zE^0(4)F = -zE^0(1)F - zE^0(2)F - zE^0(3)F$$

Cancelling through by F and substituting in values of z, and the values of E^0 (from Fig. 3.3) for steps (1) to (3) gives:

$$-zE^0(4) = -zE^0(1) - zE^0(2) - zE^0(3)$$

$$-(4 \times E^0(4)) = -(1.00) - (0.59) - (2 \times 2.00)$$

$$E^0(4) = +1.40 \text{ V}$$

This confirms the value given in Fig. 3.3. Using this method, data from a potential diagram can be used to deduce the standard reduction potentials for steps not specifically designated on the diagram, for example, E^0 for the reduction of WO_3 to WO_2 in Fig. 3.1.

A change from acidic to basic solution (or an even less drastic change in pH) can result in very significant changes in the potential diagram for a given element. Firstly, the species represented in the diagram may change as a function of pH. Secondly, values of E^0 corresponding to a particular redox couple may be pH dependent. To illustrate the first point, compare the potential diagrams for tungsten in solutions at pH 0 (Fig. 3.1) and pH 14 (Fig. 3.4). Tungsten(VI) is in the form of WO_3 at pH 0, but as $[WO_4]^{2-}$ at pH 14.

Exercise: write equations to show the half-reactions to which the W(VI) to W(IV), W(IV) to W(0), and W(VI) to W(0) steps in Fig. 3.4 refer.

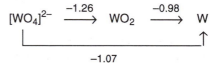

Fig. 3.4. Potential diagram for tungsten in basic solution (pH 14); E^0 values in V.

The values of the reduction potentials for equilibria which *involve* H^+ (in acidic solution) or $[OH]^-$ (in alkaline solution) are sensitive to changes in pH. The relationship between E and the concentration of solution species for a general half-equation 3.5 is given in Eqn 3.6.

$$m(Ox) + ze^- \rightleftharpoons n(Red) \tag{3.5}$$

$$E = E^\circ - \frac{RT}{zF} \ln \frac{[\text{reduced species}]^n}{[\text{oxidized species}]^m} \qquad \textit{Nernst equation} \tag{3.6}$$

$R = 8.314 \ J \ K^{-1} \ mol^{-1}$
T = temperature in K (usually 298 K)
z = number of electrons transferred
F = 96 485 C

In the specific case of, for example, reaction 3.7, the Nernst equation has the form of Eqn 3.8.

$$[Cr_2O_7]^{2-}(aq) + 14H^+(aq) + 6e^- \rightleftharpoons 2Cr^{3+}(aq) + 7H_2O(l) \tag{3.7}$$

$$E = E^\circ - \frac{RT}{zF} \ln \frac{[Cr^{3+}]^2}{[Cr_2O_7^{2-}][H^+]^{14}} \tag{3.8}$$

The Nernst equation should be written in terms of the activities of the species present, but at *low concentrations*, activities may be approximated to concentrations.

Water is not included in the equation since its activity is taken to be unity. From Eqn 3.8, it is clear that the reduction potential will be pH dependent (pH = $-\lg [H^+]$).

In the remainder of this chapter, we discuss the species present in aqueous solutions of second and third row *d*-block metals in different oxidation states. Lanthanum is excluded (see page 1).

3.3 Group 3: yttrium

Yttrium is often considered with the *f*-block elements, and its chemistry is almost invariably concerned with the +3 oxidation state. Aqua ions with high coordination numbers are known, and both $[Y(H_2O)_8]^{3+}$ and $[Y(H_2O)_9]^{3+}$ have been structurally characterized (X-ray or neutron diffraction) in a number of different salts. The structure of $[Y(H_2O)_9]^{3+}$ is shown in Fig. 3.5, where

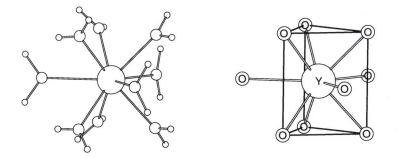

Fig. 3.5. The $[Y(H_2O)_9]^{3+}$ cation has been characterized in the solid state by X-ray diffraction; the O atoms define a tricapped trigonal prism around the Y(III) centre. Each 'cap' is defined as shown for the monocapped trigonal prism (Fig. 2.2 and associated text).

the trigonal prismatic coordination shell of the nine oxygen donor atoms has been emphasized.

3.4 Group 4: zirconium and hafnium

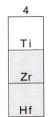

In aqueous solution, only the +4 oxidation states of zirconium and hafnium are stable. It is important to note that the M^{4+} ion is not present and thus, it is more satisfactory to refer to the M(IV) state, rather than M^{4+}, even though one may find potential diagrams written in the form if Fig. 3.6. The aqua ions of M(IV) hydrolyse, but the species present are not simple and are not necessarily mononuclear, their nature being dependent upon exact conditions such as concentration and counter-ion.

In basic solution, the M(IV) state is present as $MO(OH)_2$, and reduction to M(0) is according to half-reaction 3.9 with reduction potentials of -2.36 V for M = Zr and -2.50 V for Hf.

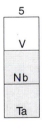

$$MO(OH)_2 + 4H_2O + 4e^- \rightleftharpoons M + 4[OH]^- \tag{3.9}$$

Fig. 3.6. Potential diagrams for Zr and Hf; see comment in text.

3.5 Group 5: niobium and tantalum

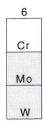

The aqueous solution chemistry of niobium and tantalum does not involve simple aqua ions for the lower oxidation states, but, rather, higher nuclearity species. Additionally, the +5 oxidation state is dominant in the chemistry of these metals, and solution redox chemistry is not a major aspect in the chemistry of niobium and tantalum; compare this to the chemistry of vanadium, the first member of group 5. As for zirconium and hafnium, potential diagrams for niobium and tantalum can be misleading in terms of the species present in solution. Although a value of $E^O = -1.10$ V for the reduction of Nb^{3+} to Nb is found in tables of potential data, it is likely that more complex species are present and one should regard this as Nb(III) rather than the Nb^{3+} ion.

3.6 Group 6: molybdenum and tungsten

We have already mentioned some of the redox chemistry of tungsten (Section 3.2), and have noted that although Cr(VI) is a strong oxidizing agent in acidic solution, analogous chemistry for Mo(VI) and W(VI) is not important. The metal(VI) oxides (MO_3) dissolve in alkaline solutions to give oxoanions; however, the condensation (which is pH dependent) of $[MO_4]^{2-}$ (M = Mo or W) oxoanions complicates the picture so far as solution M(VI) species are concerned, as half-equation 3.10 illustrates. We expand on this statement in Chapter 8.

$$[H_3Mo_7O_{24}]^{3-} + 45H^+ + 42e^- \rightleftharpoons 7Mo + 24H_2O \qquad E^O = +0.08 \text{ V} \tag{3.10}$$

Molybdenum(VI) oxide dissolves in acids to give cationic species which include $[Mo(O)_2(H_2O)_4]^{2+}$ (**3.4**) and $[Mo(OH)_4(H_2O)_2]^{2+}$. Thus, reduction equilibrium 3.11, which is quoted in standard tables of redox potentials, is not as simple as it may appear in terms of the species involved.

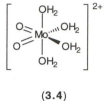

(**3.4**)

$$MoO_3 + 6H^+ + 6e^- \rightleftharpoons Mo + 3H_2O \qquad E^\circ = +0.07\ V \qquad (3.11)$$

For the lower oxidation states, the extent to which solution species have been well characterized is greater for Mo than for W, and we therefore confine our discussion to the former. The hexaaqua ion of Mo^{3+} is very air-sensitive; it has been isolated in the yellow, 'caesium molybdenum alum', $CsMo(SO_4)_2 \cdot 12H_2O$, the crystal structure of which confirms the presence of an octahedral $[Mo(H_2O)_6]^{3+}$ ion. No such simple aqua ion is known for Mo(II); representative of Mo(II) is the dinuclear cation $[Mo_2(H_2O)_8]^{4+}$ which is formed in acidic solution from $K_4[Mo_2(SO_4)_4]$. The structure of $[Mo_2(H_2O)_8]^{4+}$ (**3.5**) has been proposed on the basis of its electronic spectrum which indicates the presence of the molybdenum-molybdenum quadruple bond; (see Chapter 6 for a discussion of metal-metal multiple bonds). The $[Mo(H_2O)_6]^{3+}$ ion is not unique for the aqua species of Mo(III); the blue-green hydroxy-bridged $[Mo_2(\mu\text{-}OH)_2(H_2O)_8]^{4+}$ (**3.6**) formally contains the $\{Mo_2\}^{6+}$ core and is formed in aqueous solution when $Na_2[MoO_4]$ is reduced with zinc amalgam.

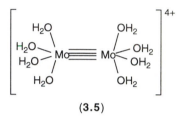

(**3.5**)

Di- and trinuclear species are favoured for aqua anions of Mo(IV) and Mo(V). Molybdenum(IV) is stabilized in acidic solution as the red cation $[Mo_3O_4(H_2O)_9]^{4+}$; the oxidation state can be assigned by considering the presence of an $\{Mo_3\}^{12+}$ unit, although, of course, this is a pure formalism. $[Mo_3O_4(H_2O)_9]^{4+}$ may be produced either by reduction of $Na_2[MoO_4]$ or oxidation of $[Mo_2(H_2O)_8]^{4+}$ under appropriate conditions, and its structure is shown in Fig. 3.7. One oxygen atom is triply bridging (μ_3-mode) and three

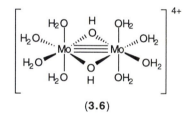

(**3.6**)

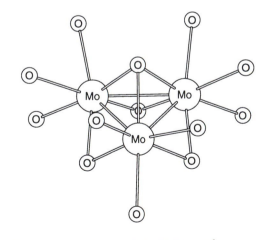

Fig. 3.7. The structure of $[Mo_3(\mu_3\text{-}O)(\mu\text{-}O)_3(H_2O)_9]^{4+}$, determined by X-ray diffraction for the $[4\text{-}MeC_6H_4SO_3]^-$ salt. Hydrogen atoms are omitted for clarity; each *terminal* O represents a water ligand.

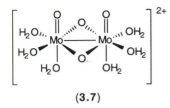

(3.7)

are doubly bridging, and the formula is more appropriately written as $[Mo_3(\mu_3\text{-}O)(\mu\text{-}O)_3(H_2O)_9]^{4+}$. The Mo–Mo distances of 247 to 249 pm indicate bonding interactions.

When the pyridinium salt of $[Mo(O)Cl_5]^{2-}$ (i.e. Mo(V)) is dissolved in aqueous acid, the solution species present is the yellow $[Mo_2(O)_2(\mu\text{-}O)_2(H_2O)_6]^{2+}$, and this is also produced in solution when $Na_2[MoO_4]$ is reduced by hydrazinium chloride. The proposed structure for the cation $[Mo_2(O)_2(\mu\text{-}O)_2(H_2O)_6]^{2+}$, **3.7**, includes an Mo–Mo single bond, consistent with the observed diamagnetism of the ion.

3.7 Group 7: technetium and rhenium

For both technetium and rhenium, oxo complexes (i.e. Tc=O or Re=O bond formation) are important in aqueous solution chemistry. Probably the most common starting materials in technetium chemistry are pertechnetate salts containing $[TcO_4]^-$, (Fig. 3.8) in which technetium is in its highest oxidation state, Tc(VII). This, and the isostructural $[ReO_4]^-$ ion, are by no means as good oxidizing agents as $[MnO_4]^-$ as Fig. 3.9 illustrates. The ions $[TcO_4]^-$ and $[ReO_4]^-$ are generated when Tc_2O_7 and Re_2O_7, respectively, are dissolved in aqueous solution.

For the lower oxidation states, aqua ions such as $[Tc(H_2O)_6]^{3+}$ have not been characterized, although the technetium(I) aqua ion $[Tc(H_2O)_3(CO)_3]^+$ is a precursor for the development of ^{99m}Tc labelled pharmaceuticals.

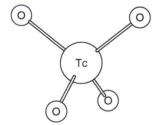

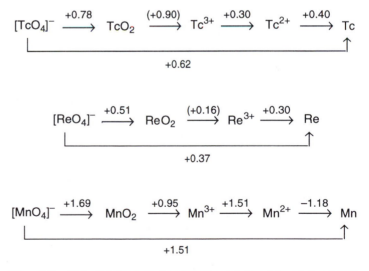

Fig. 3.8. The tetrahedral structure of the $[TcO_4]^-$ anion.

Fig. 3.9. Potential diagrams for group 7 metals in acidic solution (pH 0); E^O values in V.

3.8 Group 8: ruthenium and osmium

In the tetraoxides RuO_4 and OsO_4, ruthenium and osmium exhibit an oxidation state of +8, the highest oxidation state attainable and comparable with that of xenon in XeO_4. RuO_4 and OsO_4 are toxic and volatile compounds, with OsO_4 being more stable than RuO_4 with respect to reduction. The chemistry of these species contrasts greatly with that of the first member of group 8, iron, where the highest common oxidation state is +6, observed in $[FeO_4]^{2-}$.

In aqueous alkali, RuO_4 is reduced to $[RuO_4]^-$, and in concentrated alkaline solution, further reduction takes place to give $[RuO_4]^{2-}$ (Eqn 3.12).

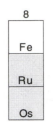

Caution! RuO_4 reacts explosively with many organic solvents.

$$2[RuO_4]^-(aq) + 2[OH]^-(aq) \rightleftharpoons 2[RuO_4]^{2-}(aq) + H_2O(l) + {}^1/_2O_2(g) \qquad (3.12)$$

The rate of formation of $[RuO_4]^{2-}$ in this system increases as $[OH^-]$ increases. Acidification of these solutions results in disproportionation of the ruthenium-containing species. The $[RuO_4]^-$ and $[RuO_4]^{2-}$ ions can be stabilized in solution, each under appropriate pH conditions and $[RuO_4]^{2-}$ in a non-reducing environment. Salts of $[RuO_4]^-$ can be isolated, e.g. $K[RuO_4]$; the $[RuO_4]^-$ anion has a compressed tetrahedral structure. In the case of '$K_2[RuO_4]$', the isolated crystalline monohydrate has been shown to be $K_2[RuO_3(OH)_2]$, for which structure **3.8** has been confirmed for the anion. A related example is '$Ba[RuO_4]\cdot H_2O$' which has been shown to contain $[RuO_3(OH)_2]^{2-}$. Both $[RuO_4]^-$ and $[RuO_4]^{2-}$ are powerful oxidizing agents.

(3.8)

Osmium tetraoxide (also called osmic acid) dissolves in aqueous alkali *without* reduction to give $[OsO_4(OH)_2]^{2-}$. Infrared spectroscopic data are consistent with the *cis*-isomer shown in structure **3.9**; crystalline salts (e.g. $Na_2[OsO_4(OH)_2]\cdot 2H_2O$) can be isolated. Reduction of $[OsO_4(OH)_2]^{2-}$ in ethanolic KOH leads to *trans*-$[OsO_2(OH)_4]^{2-}$.

Aqua ions of the $[M(H_2O)_6]^{n+}$ type are observed for Ru^{2+} and Ru^{3+}, but have not been clearly established for Os^{n+} ($n = 2$ or 3). The formation in solution of $[Ru(H_2O)_6]^{n+}$ ($n = 2$ or 3) is not as straight forward as that of the corresponding $[Fe(H_2O)_6]^{n+}$ solution species. The aqua ion $[Ru(H_2O)_6]^{2+}$ may be produced by reduction of RuO_4 in aqueous solution using lead as reductant. It has been isolated as the tosylate salt, $[Ru(H_2O)_6][tos]_2$; $[tos]^- =$ **3.10**. Oxidation with O_2 leads to the formation of $[Ru(H_2O)_6]^{3+}$, and this ion can be isolated as the salt $[Ru(H_2O)_6][tos]_3\cdot 3H_2O$. Both tosylate salts have been characterized by X-ray diffraction, and the octahedral structures of the cations confirmed; a comparison of the solid state structures shows that in going from $[Ru(H_2O)_6]^{2+}$ to $[Ru(H_2O)_6]^{3+}$, the Ru–O bond lengths shorten from 212.2 to 202.9 pm. The aqua ions $[Ru(H_2O)_6]^{2+}$ and $[Ru(H_2O)_6]^{3+}$ are, respectively, present in Tutton salts of formula $M_2Ru(SO_4)_2\cdot 6H_2O$ (M = Rb, NH_4) and in the alum $CsRu(SO_4)_2\cdot 12H_2O$. The latter is analogous to the presence of $[Fe(H_2O)_6]^{3+}$ in the alum $KFe(SO_4)_2\cdot 12H_2O$. Loss of a proton from $[Ru(H_2O)_6]^{3+}$ occurs according to equilibrium 3.13, and a pK_a value of 2.4 has been estimated from electronic

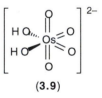

(3.9)

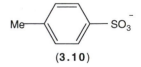

(3.10)

spectroscopic data; this compares to a value of $pK_a = 2.0$ for $[Fe(H_2O)_6]^{3+}$ (Eqn 3.1)

$$[Ru(H_2O)_6]^{3+}(aq) + H_2O(l) \rightleftharpoons [Ru(H_2O)_5(OH)]^{2+}(aq) + [H_3O]^+(aq) \qquad (3.13)$$

EXAFS = extended X-ray absorption fine structure

The nature of the aqua ion of Ru(IV) (formed by electrochemical oxidation of $[Ru(H_2O)_6]^{2+}$) has been a point of debate for many years, but the presence of multinuclear species is expected. Oxygen-17 NMR spectroscopic data are consistent with the formation of $[Ru_4O_6(H_2O)_{12}]^{4+}$, although the degree of protonation is pH dependent. Two possible structures have been proposed and are shown in Fig. 3.10; recent EXAFS studies support the adamantanoid model.

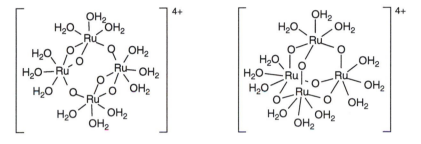

Fig. 3.10. Two proposed, alternative structures for the ruthenium(IV) aqua ion $[Ru_4O_6(H_2O)_{12}]^{4+}$; left, edge-bridged Ru_4-rectangle, and right, adamantanoid structure.

3.9 Group 9: rhodium and iridium

The oxidation states that predominate in the chemistries of rhodium and iridium are Rh(III), and Ir(III) and Ir(IV) respectively.

The formation of $[Rh(H_2O)_6]^{3+}$ in aqueous solution can be achieved in the presence of perchloric acid; crystals of $Rh(ClO_4)_3 \cdot 6H_2O$ contain the octahedral $[Rh(H_2O)_6]^{3+}$ ion. In aqueous solution, $[Rh(H_2O)_6]^{3+}$ hydrolyses according to Eqn 3.14.

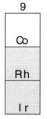

$$[Rh(H_2O)_6]^{3+}(aq) + H_2O(l) \rightleftharpoons [Rh(H_2O)_5(OH)]^{2+}(aq) + [H_3O]^+(aq)$$

$$pK_a = 3.33 \qquad (3.17)$$

The iridium analogue, $[Ir(H_2O)_6]^{3+}$, can be prepared by hydrolysing the iridium(IV) salt $[NH_4]_2[IrCl_6]$, but the product is air sensitive and the aqua ion has not been fully characterized. However, $[Ir(H_2O)_6]^{3+}$ is present in $CsIr(SO_4)_2 \cdot 12H_2O$, and the corresponding alum $CsRh(SO_4)_2 \cdot 12H_2O$ contains $[Rh(H_2O)_6]^{3+}$.

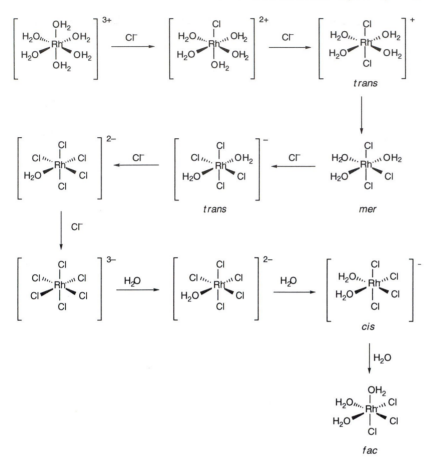

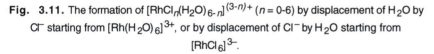

Fig. 3.11. The formation of $[RhCl_n(H_2O)_{6-n}]^{(3-n)+}$ (n = 0-6) by displacement of H_2O by Cl^- starting from $[Rh(H_2O)_6]^{3+}$, or by displacement of Cl^- by H_2O starting from $[RhCl_6]^{3-}$.

The ions $[M(H_2O)_6]^{2+}$ (M = Rh or Ir) are not formed by reduction of the corresponding M(III) aqua ions. However, reduction of $[RhCl(H_2O)_5]^{2+}$ by $[Cr(H_2O)_6]^{2+}$ leads to the formation of $[Rh_2(H_2O)_{10}]^{4+}$, frequently written as $[Rh_2]^{4+}(aq)$. Full characterization of this diamagnetic aqua species has not been achieved. The chloro species $[RhCl(H_2O)_5]^{2+}$ from which $[Rh_2]^{4+}(aq)$ is prepared is one member of a series of such ions of general formula $[RhCl_n(H_2O)_{6-n}]^{(3-n)+}$ where n = 0-6. These ions can be prepared in solution either by displacement of H_2O by chloride starting from $[Rh(H_2O)_6]^{3+}$, or by displacement of chloride by H_2O starting from $[RhCl_6]^{3-}$. The system has been well studied and the reactions shown in Fig.3.11 illustrate how the preparative routes used influence the isomers produced; the *trans* effect of the chloride ligand is greater than that of H_2O. Some members of the series of Ir(IV) complexes $[IrCl_n(H_2O)_{6-n}]^{(4-n)+}$ are known; for n = 2, the *cis* and *trans*-isomers, and for n = 3, both the *mer* and *fac*-isomers have been characterized; they are prepared by the oxidation of the analogous iridium(III)

complexes. The crystallographically determined structure of *trans-*[IrCl$_2$(H$_2$O)$_4$]$^+$ is shown in Fig. 3.12.

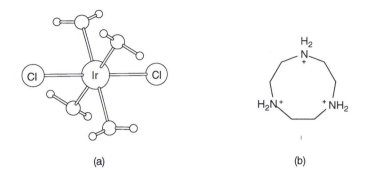

(a)   (b)

Fig. 3.12. (a) The structure of *trans*-[IrCl$_2$(H$_2$O)$_4$]$^+$ has been determined (X-ray diffraction) in the solid state for the compound [C$_6$H$_{18}$N$_3$][IrCl$_2$(H$_2$O)$_4$][SO$_4$]$_2$, the [C$_6$H$_{18}$N$_3$]$^{3+}$ cation of which is shown in (b).

3.10 Group 10: palladium and platinum

Aqua ions of palladium(II) and platinum(II) are known in solution but have not been isolated in salts. [Pd(H$_2$O)$_4$]$^{2+}$ can be produced in strongly acidic solutions (dissolving PdO in perchloric acid), and [Pt(H$_2$O)$_4$]$^{2+}$, by treating aqueous [PtCl$_4$]$^{2-}$ with silver(I) perchlorate. Both [Pd(H$_2$O)$_4$]$^{2+}$ and [Pt(H$_2$O)$_4$]$^{2+}$ have been the subjects of detailed kinetic studies, and substitution at the platinum(II) centre is more difficult than in [Pd(H$_2$O)$_4$]$^{2+}$. As equations 3.15 to 3.17 show, both Pd^{2+}(aq) and Pt^{2+}(aq) are significantly better oxidizing agents than Ni^{2+}(aq).

$$Ni^{2+}(aq) + 2e^- \rightleftharpoons Ni(s) \qquad\qquad E^o = -0.25 \text{ V} \qquad (3.15)$$

$$Pd^{2+}(aq) + 2e^- \rightleftharpoons Pd(s) \qquad\qquad E^o = +0.99 \text{ V} \qquad (3.16)$$

$$Pt^{2+}(aq) + 2e^- \rightleftharpoons Pt(s) \qquad\qquad E^o \approx +1.2 \text{ V} \qquad (3.17)$$

3.11 Group 11: silver and gold

In group 11, the situation as regards oxidation state stabilities is rather different than that for earlier groups in the *d*-block. For group 11 metals, the important oxidation states are copper(II), copper(I), silver(I), gold(I) and gold(III). Relativistic effects are considered to be a crucial factor in the observed stability of the +3 oxidation state for gold. Equations 3.18 to 3.22 show the standard reduction potentials for silver and gold ions in solution; the half-reactions are arranged in order of E^o values, with Ag^{2+} shown to be a very powerful oxidizing agent. Silver(II) can be formed in perchloric acid

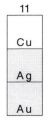

Relativistic effects: see page 52.

solution either by oxidation of Ag^+ using O_3 as oxidant, or dissolution of AgO. However, once formed, the tendency for Ag^{2+}(aq) to be reduced is high.

$$Ag^+(aq) + e^- \rightleftharpoons Ag(s) \qquad\qquad E^0 = +0.80 \text{ V} \qquad (3.18)$$

$$[AuCl_4]^-(aq) + 3e^- \rightleftharpoons Au(s) + 4Cl^-(aq)$$

$$E^0_{(at\ [Cl^-]\ =1\ M)} = +1.00 \text{ V} \qquad (3.19)$$

$$Au^{3+}(aq) + 2e^- \rightleftharpoons Au^+(aq) \qquad E^0 = +1.40 \text{ V} \qquad (3.20)$$

$$Au^+(aq) + e^- \rightleftharpoons Au(s) \qquad\qquad E^0 = +1.69 \text{ V} \qquad (3.21)$$

$$Ag^{2+}(aq) + e^- \rightleftharpoons Ag^+(aq) \qquad E^0 = +1.98 \text{ V} \qquad (3.22)$$

Although readily found in tables of data, some reduction potential data should be viewed with caution; for example, in the presence of chloride ion, Au(III) forms a complex ion in solution (e.g. $[AuCl_4]^-$, **3.11**, Eqn 3.19) and is *not* present as an aqua species. Silver(I) forms an aqua ion, but the H_2O molecules are readily displaced by other ligands. Formation of a complex ion invariably alters the reduction potential of, for example, the Ag^+/Ag couple. Aqua ions of silver and gold cations do not occur in solid state salts.

Whereas copper(I) disproportionates in aqueous solution according to Eqn 3.23, silver(I) does not undergo a similar reaction. Gold(I) is unstable with respect to disproportionation, but by reaction 3.24 rather than in an analogous manner to copper(I).

(**3.11**)

$$2Cu^+(aq) \rightleftharpoons Cu^{2+}(aq) + Cu(s) \qquad (3.23)$$

$$3Au^+(aq) \rightleftharpoons Au^{3+}(aq) + 2Au(s) \qquad (3.24)$$

3.12 Group 12: cadmium and mercury

As for zinc, the +2 oxidation state is the most important for cadmium. In aqueous solution, the $[Cd(H_2O)_6]^{2+}$ ion is present, but it readily loses H^+ (Eqn 3.25). In solutions where the concentration of Cd^{2+} ions is high, a dinuclear species, $[Cd_2(OH)]^{3+}$(aq), forms.

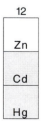

$$[Cd(H_2O)_6]^{3+}(aq) + H_2O(l) \rightleftharpoons [Cd(H_2O)_5(OH)]^{2+}(aq) + [H_3O]^+(aq) \qquad (3.25)$$

In contrast to zinc which dissolves in excess alkali to give $[Zn(OH)_4]^{2-}$, cadmium hydroxide, $Cd(OH)_2$, is insoluble in mildly basic solutions.

Mercury and cadmium in the
Weston standard cell: see page 8.

For mercury, both Hg(II) and Hg(I) are important, the latter being encountered as Hg_2^{2+}. For example, the standard calomel electrode (a standard reference electrode) consists of a platinum wire, dipping into mercury which is in contact with Hg_2Cl_2, the electrode being immersed in 1 M KCl solution (Eqn 3.26). Note that the reduction potential of the Hg_2Cl_2 / Hg, $2Cl^-$ couple varies (like any cell) with cell conditions, and that, for example, the *standard* (Eqn 3.26) and *saturated* calomel electrodes are not the same as each other.

$$Hg_2Cl_2 + 2e^- \rightleftharpoons 2Hg + 2Cl^- \qquad\qquad E^0 = +0.268\ V \qquad (3.26)$$

(in 1 M KCl solution)

$$Hg_2^{2+} \rightleftharpoons Hg^{2+} + Hg \qquad\qquad E^0_{cell} = -0.14\ V \qquad (3.27)$$

The potential diagrams in Fig. 3.13 summarize the redox chemistry of cadmium and mercury in acidic and basic solutions. Looking at the E^0 values for mercury species in acidic solution, it is apparent that the disproportionation of Hg(I) to Hg and Hg(II) (Eqn 3.27) has a *small* and *positive* ΔG^0 value.

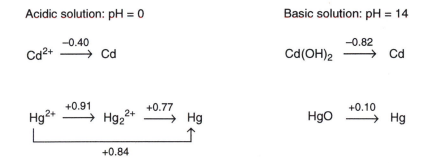

Fig. 3.13. Potential diagrams for cadmium and mercury; E^0 values in V.

4 Coordination complexes: structure

4.1 Coordination numbers and structures

In considering the second and third row metals in the *d*-block, it is hard to generalize about trends in coordination numbers. The aim of this section is to introduce the ranges of coordination numbers that are observed for the heavier *d*-block metals and to give selected examples of complexes. Table 4.1 gives common ligand polyhedra, but several points must be noted:

- for coordination numbers of five or eight, the energy difference between different possible structures are often small; fluxional behaviour in solution may be observed, and the structure of, for example, an 8-coordinate complex anion, may be cation-dependent.
- distortions from an ideal polyhedron are often encountered due to electronic and/or steric effects, an example being the restriction that the bite angle of a chelating ligand may impose.
- in dinuclear complexes, it is usually appropriate to describe the coordination geometries with respect to the individual metal centres.

Note: The reader should be aware that discussions exclude organometallic species.

Coordination numbers: see also Section 2.3

4.2 Yttrium

The coordination chemistry of yttrium is that of the Y^{3+} ion, and complexes exhibit a range of coordination numbers, for example *trans*-$[YCl_4(thf)_2]^-$

See page 87 for ligand abbreviations and structures.

Table 4.1 Ligand polyhedra (defined by the positions of the coordinated donor atoms).

Coordination number	Arrangement of donor atoms around metal centre	Less common arrangements
2	Linear	
3	Trigonal planar	T-shaped; trigonal pyramidal
4	Tetrahedral; square planar	
5	Trigonal bipyramidal; square-based pyramidal	
6	Octahedral	Trigonal prismatic
7	Pentagonal bipyramidal	Monocapped trigonal prismatic
8	Dodecahedral; square antiprismatic; hexagonal bipyramidal	Cube; bicapped trigonal prismatic
9	Tricapped trigonal prismatic	

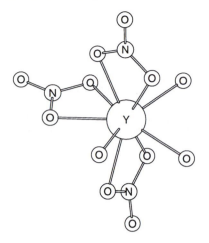

Fig. 4.1. The structure of [Y(NO$_3$)$_3$(H$_2$O)$_3$] determined by X-ray diffraction; H atoms are omitted.

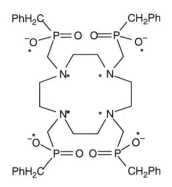

Fig. 4.2. An example of a macrocyclic ligand with pendant arms; the eight donor sites are marked by asterisks. The yttrium(III) complex with this ligand has a square antiprismatic arrangement of donor atoms.

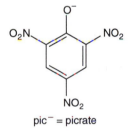

pic$^-$ = picrate

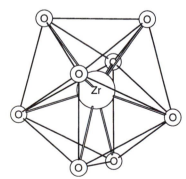

Fig. 4.3. The dodecahedral arrangement of oxygen donor atoms around Zr(IV) in [Zr(ox)$_4$]$^{4-}$.

(octahedral), *trans*-[YCl$_2$(thf)$_5$]$^+$ (pentagonal bipyramidal), [Y(pic)(H$_2$O)$_7$]$^{2+}$ (eight-coordinate), [Y(H$_2$O)$_9$]$^{3+}$ (tricapped trigonal prism, see Fig. 3.5). An example of a *non-regular* nine-coordinate geometry is provided by the complex [Y(NO$_3$)$_3$(H$_2$O)$_3$], Fig. 4.1. The arrangement of the donor atoms in a polydentate ligand clearly influences the preference for a particular coordination geometry. An example is shown in Fig. 4.2; this ligand coordinates to Y^{3+} through four *N*- and four *O*-centres to give a square antiprismatic coordination sphere about the metal ion.

Bulky amido ligands such as [N(SiMe$_3$)$_2$]$^-$ often feature in complexes with *low* coordination numbers. In [Y{N(SiMe$_3$)$_2$}$_3$], the metal centre is in a trigonal pyramidal environment ($\angle$N-Y-N = 114.6°). Such a geometry is also seen in [Sc{N(SiMe$_3$)$_2$}$_3$] and some related lanthanoid complexes, but crystal packing effects in the solid state are likely to be responsible for the deviation from planarity. This is supported by the fact that the gas phase structure of [Sc{N(SiMe$_3$)$_2$}$_3$] exhibits a trigonal planar coordination geometry.

4.2 Zirconium and hafnium

Coordination numbers for hafnium and zirconium generally range from four to eight. Formulae can be deceptive: for example, in the solid state, MF$_4$ (M = Hf or Zr) adopts a structure in which each metal centre is eight-coordinate (square antiprismatic), while in MCl$_4$, the metal centres are octahedrally sited. High coordination numbers are also observed in a range of complexes including [ZrF$_7$]$^{3-}$ (seven-coordinate, Fig. 2.2), [Zr(NO$_3$)$_4$] and [Zr(ox)$_4$]$^{4-}$ (eight-coordinate, dodecahedral, Fig. 4.3), [Zr(acac)$_4$] (eight-coordinate, square antiprismatic, Fig. 4.4). Figure 4.5 shows the structure of

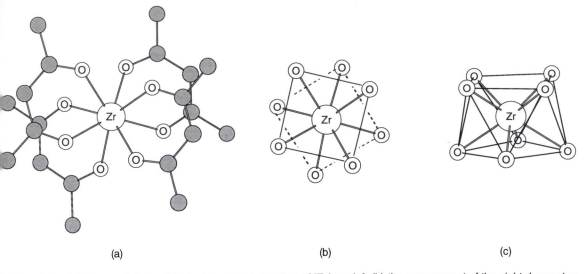

Fig. 4.4. (a) The crystallographically determined structure of [Zr(acac)$_4$]; (b) the arrangement of the eight donor atoms, four lie above the metal centre, and four below; (c) another view of the coordination sphere which emphasizes the square antiprismatic polyhedron.

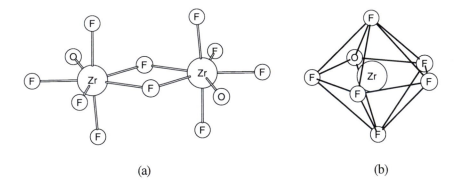

Fig. 4.5. (a) The structure of [Zr$_2$(μ-F)$_2$F$_8$(H$_2$O)$_2$]$^{2-}$, determined by X-ray diffraction for the [NMe$_4$]$^+$ salt; (b) each Zr(IV) centre is seven-coordinate with a pentagonal bipyramidal structure.

[Zr$_2$F$_8$(μ-F)$_2$(H$_2$O)$_2$]$^{2-}$ in which each Zr(IV) centre is in a pentagonal bipyramidal environment; in [Zr$_2$(μ-N$_3$)$_2$Cl$_8$]$^{2-}$, each metal centre is octahedrally coordinated, being bonded to four chloro and two bridging azido ligands. Lower coordination numbers are observed with appropriate ligands, e.g. amido groups; for example five-coordination in [Zr$_2$(NMe$_2$)$_6$(μ-NMe$_2$)$_2$] (Fig. 4.6a), and tetrahedral arrangements in [M(NPh$_2$)$_4$] and [M{N(SiMe$_3$)$_2$}$_3$Cl] (M = Hf or Zr, Fig. 4.6b).

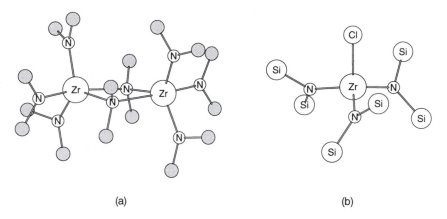

(a)   (b)

Fig. 4.6. The structures of (a) [Zr$_2$(NMe$_2$)$_6$(μ-NMe$_2$)$_2$] and (b) [Zr{N(SiMe$_3$)$_2$}$_3$Cl] determined by X-ray diffraction.

4.3 Niobium and tantalum

Just as for zirconium and hafnium, there are similarities between the stereochemistries of niobium and tantalum, with coordination numbers of six, seven or eight being prevalent. Five-coordinate complexes include [Ta(NEt$_2$)$_5$] (trigonal bipyramid), [Nb(NMe$_2$)$_5$] (square-based pyramid), and [Nb(O)Cl$_4$]$^-$ (square-based pyramid). However, the pentafluorides, (**4.1**), are tetrameric, resulting in coordination numbers of six about each metal centre, and six-coordinate metal centres are also observed in the pentachlorides which are dimeric (**4.2**). Octahedral geometries are also seen in metal(V) halo anions, e.g. [TaF$_6$]$^-$, [TaCl$_6$]$^-$ and [NbBr$_6$]$^-$, as well as in a range of other complexes including *cis*-[TaCl$_4$(PMe$_2$Ph)$_2$], *trans*-[TaCl$_4$(PEt$_3$)$_2$] and the oxo-bridged [Ta$_2$F$_{10}$(μ-O)]$^{2-}$ (Fig. 4.7a). The linear environment about the oxygen centre and the Ta-O bond lengths of 187.5 pm are indicative of multiple bond character. In [NbF$_4$(O)(H$_2$O)]$^-$ (Fig. 4.7b), it is interesting to note the distortion of the four F atoms out of the equatorial plane and away from the oxo-ligand; the O$_{oxo}$ and O$_{water}$ donor atoms can readily be distinguished by the difference in Nb–O bond lengths (Fig. 4.7b).

Seven-coordination is exemplified by [NbCl$_4$(PMe$_3$)$_3$], and, like a number of similar species, the polyhedron defined by the donor atoms is described as capped octahedral (Fig. 4.8a) or, in some cases, 'irregular'. Pentagonal bipyramidal complexes include [Nb(O)(H$_2$O)$_2$(ox)$_2$]$^-$ (Fig. 4.8b) and [Nb(O)(ox)$_3$]$^{3-}$ (in which the oxo ligand occupies an axial site). Eight-coordinate structures are less common than six- or seven-coordination, and an example is [Nb(ox)$_4$]$^{4-}$ which possesses a square antiprismatic arrangement of donor atoms.

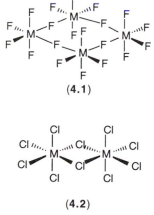

(**4.1**)

(**4.2**)

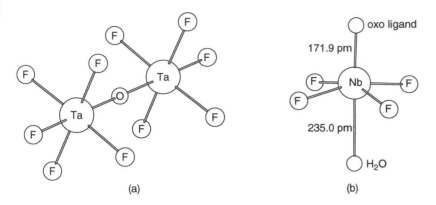

(a)   (b)

Fig. 4.7. The crystallographically determined structures of (a) $[Ta_2F_{10}(\mu\text{-O})]^{2-}$ (determined for the $[Et_4N]^+$ salt) and (b) $[NbF_4(O)(H_2O)]^-$ (determined for the $[Et_3NH]^+$ salt).

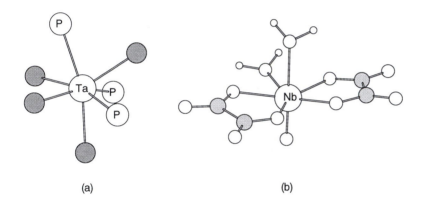

(a)   (b)

Fig. 4.8. (a) The monocapped octahedral structure of $[TaCl_4(PMe_3)_3]$ (determined by X-ray diffraction) and (b) the pentagonal bipyramidal structure of $[Nb(O)(H_2O)_2(ox)_2]^-$ (determined by neutron diffraction).

4.4 Molybdenum and tungsten

This section focuses on mononuclear species of Mo and W; metal-metal bonded and cluster species are discussed in later chapters. As we have already seen, molecular stoichiometry in mononuclear compounds such as halides does not necessarily provide information about coordination number in the solid state. The halides MX_3 form extended lattices or chains; for example, $MoBr_3$ consists of face-sharing octahedra, adjacent metal centres being linked through three halide bridges. Part of this structural motif is seen in $[M_2Cl_9]^{3-}$ and $[M_2Br_9]^{3-}$ (M = Mo or W) (**4.3**), although the metal-metal separation and magnetic data indicate that the metal centres interact significantly.

Unambiguous octahedral coordination is observed in, for example, WF_6, MoF_6, $[WF_6]^-$, $[WCl_6]^{2-}$ and $[MoCl_6]^{3-}$, in the oxo-complexes $[MoF_5(O)]^{2-}$, $[WF_5(O)]^-$ and *cis*-$[MoF_4(O)_2]^{2-}$, and in the amido and alkoxy

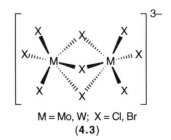

M = Mo, W; X = Cl, Br
(4.3)

complexes [W(NMe$_2$)$_6$] and [W(OMe)$_6$]; note the varied oxidation states of the metals in these octahedral species. The pentafluorides of molybdenum and tungsten form tetramers which are isostructural with those of niobium and tantalum (**4.1**). Similarly, MoCl$_5$ and WCl$_5$ are dimers with structures analogous to those of NbCl$_5$ and TaCl$_5$ (**4.2**).

A range of molybdenum and tungsten species containing oxo (M=O) or thio (M=S) ligands adopt square-based pyramidal structures, with an oxo or thio ligand occupying the apical site. One example is [WCl$_4$(O)]$^-$; however, while the neutral tungsten(VI) compound [WCl$_4$(O)] has a discrete molecular structure in the gas phase, the solid state structure consists of chains in which octahedral W centres are bridged by oxo-ligands. The structure of [Mo(S)(S$_4$)$_2$]$^{2-}$ (less informatively written as [MoS$_9$]$^{2-}$) is shown in Fig. 4.9a; besides illustrating five-coordination, it also exemplifies one of a range of species involving polysulfido ligands, in this case [S$_4$]$^{2-}$, which may act as chelating and/or bridging ligands.

Peroxo complexes of molybdenum are of particular interest because of their catalytic behaviour with respect to the epoxidation of alkenes. They also provide examples of molybdenum complexes with high-coordination numbers. In the descriptions that follow, each η^2-[O$_2$]$^{2-}$ ligand is assumed to occupy *two* coordination sites; as with η^2-alkene ligands, some ambiguity over this description may arise. Seven-coordinate structures are adopted by a number of peroxo complexes including [M(O$_2$)$_2$(O)(ox)]$^{2-}$ (M = Mo, W, Fig. 4.9b). In [Mo(O$_2$)$_4$]$^{2-}$, the eight oxygen atoms form a dodecahedral (see Fig. 4.3) array, and studies of the electronic structure of this ion indicate that strong covalent interaction between the metal 4d and peroxo σ_g(p) orbitals leads to significant weakening of the O–O bond. This weakening is greater than in the analogous chromium species.

Tetrahedral coordination is observed in ions such as [MoS$_4$]$^{2-}$, [WS$_4$]$^{2-}$, [WSe$_4$]$^{2-}$.

[O$_2$]$^{2-}$ = peroxo ligand (peroxide ion)

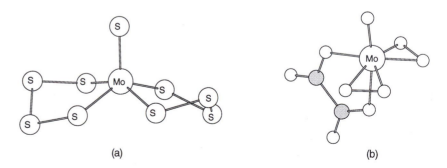

(a)   (b)

Fig. 4.9. The structures (confirmed by X-ray diffraction) of (a) [Mo(S)(S$_4$)$_2$]$^{2-}$, determined for the [NEt$_4$]$^+$ salt, and (b) [Mo(O$_2$)$_2$(O)(ox)]$^{2-}$, determined by for the potassium salt.

4.5 Technetium and rhenium

High coordination numbers are achieved for a number of hydride complexes of Re and Tc, and in $[ReF_8]^{2-}$ which has a square antiprismatic structure. The coordination sphere in $[MH_9]^{2-}$ (M = Re, Tc) is a tricapped trigonal prism (see Fig. 3.5), and this is retained in a number of derivatives such as $[ReH_7(dppe)]$. However, some hydrido complexes feature coordinated H_2, although the H–H bond may be 'stretched' and the molecule on its way to becoming two H atoms. In such cases, the H_2 ligand may be considered to occupy one coordination site. An example is $[ReH_7\{P(C_6H_4\text{-}4\text{-}Me)_3\}_2]$ in which the coordination sphere is distorted dodecahedral; compare Fig. 4.10 with the dodecahedral array in Fig. 4.3.

Ligand abbreviations: see page 87

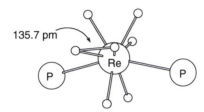

135.7 pm

Fig. 4.10 The structure (neutron diffraction) of the ReH_7P_2 core in $[ReH_7\{P(C_6H_4\text{-}4\text{-}Me)_3\}_2]$.

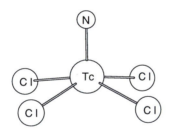

Fig. 4.11. The structure of $[TcCl_4(N)]^-$ (from an X-ray diffraction study of the $[Ph_4As]^+$ salt). The N^{3-} ligand occupies an apical site, and the Tc(VI) centre is situated above the basal plane.

Although high coordination numbers are observed in a limited number of complexes, five- and six-coordinations are more common for rhenium and technetium. Many six-coordinate complexes are known, e.g. $[TcCl_6]^{2-}$, $[ReCl_5(O)]^{2-}$, $[Re(O)Cl_4(py)]^-$, *trans*-$[Tc(O)_2(en)_2]^+$, $[Tc\{SC(NH_2)_2\}]^{3+}$, *mer*-$[TcCl_3(PMe_2Ph)_3]$ and *mer*-$[ReCl_3(NO)_2(NCMe)]$. In their higher oxidation states, Re and Tc form a range of oxo and nitrido complexes; five-coordination (which is well exemplified) shows a preference for square-based pyramidal structures with the oxo or nitrido ligand occupying the apical site, e.g. $[MCl_4(O)]^-$ (M = Re, Tc), $[TcCl_4(N)]^-$ (Fig. 4.11), $[TcBr_4(N)]^-$, $[Re(O)(C_2O_2S_2\text{-}S,S')_2]^-$ (containing the 1,2-dithiooxalate ligand) and $[Tc(N)\{S_2C_2(CN)_2\text{-}S,S'\}_2]^{2-}$ (containing the 1,2-dicyanoethenedithiolato ligand). The last two complexes are examples of five-coordinate Re(V) or

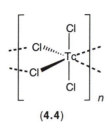

(4.4)

catechol

Tc(V) oxo or nitrido complexes in which the basal sites are occupied by a multidentate ligand. Such Tc(V) complexes have gained recent use in nuclear medicine.

As we have already seen, stoichiometry of halides in particular may not be instructive in terms of coordination numbers. The solid state structure of $ReCl_4$ contains zig-zag chains of face-sharing octahedra in which there are significant Re-Re bonding interactions; the Re centres are six-coordinate. $TcCl_4$ possesses a polymeric structure (**4.4**) with bridging Cl atoms and the Tc(IV) centres octahedrally sited. *Discrete* tetrahedral complexes include $[MO_4]^-$ and $[MO_4]^{2-}$ (M = Re, Tc; see Fig. 3.8).

4.6   Ruthenium and osmium

For the heavier group 8 metals, six-coordinate complexes are the most common. Examples of octahedral mononuclear complexes covering a range of oxidation states include $[Os(O)_2F_4]$, $[Os(O)F_5]$, $[Os(cat)_3]$ (cat^{2-} = catecholate), OsF_6, $[OsCl_6]^-$, *cis*- and *trans*-$[OsCl_2F_4]^{2-}$, $[Os(CN)_6]^{3-}$, $[Os(en)_3]^{3+}$, $[Os(bpy)_2Cl_2]$, $[Ru(O)_2(MeCO_2)_2(py)_2]$, $[RuCl(O)(py)_4]^+$, $[Ru(acac)_3]^-$, $[RuCl_5(H_2O)]^{2-}$ and $[Ru(bpy)_3]^{2+}$ (Fig. 4.12).

The complex $[Ru(bpy)_3]^{2+}$ is photoactive and has a relatively long-lived MLCT state $[Ru(bpy)_3]^{2+*}$ in which an electron has been transferred from metal to ligand. This excited species is both a better oxidizing agent and a better reducing agent than the ground state and the complex has potential application for 'water splitting', i.e. photoconversion of water to H_2 and O_2.

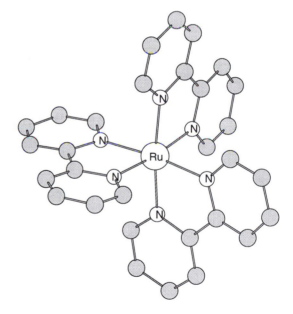

Fig.   4.12. The structure of one enantiomer of $[Ru(bpy)_3]^{2+}$ (X-ray diffraction for the $[PF_6]^-$ salt.)

In high oxidation states, (+8 being the highest attained), tetrahedral geometries are observed, e.g. in OsO_4, $[Os(O)_3N]^-$, RuO_4 and the Os(VI) complex $[Os(O)_2(S_2O_3)_2]^{2-}$ (Fig. 4.13). The addition of two electron donor ligands to, for example, OsO_4 may lead to five-coordinate complexes such as

[OsCl(O)$_4$]$^-$, [Os(4-NCpy)(O)$_4$] and [Os(4-Phpy)(O)$_4$]. These all have trigonal bipyramidal structures with the non-oxo ligand in an axial site and with the equatorial Os–O bonds bent towards the non-oxo ligand (Fig. 4.14). Five-coordinate species for lower oxidation states are also known, both with trigonal bipyramidal and square-based pyramidal structures, e.g. [OsCl$_2$(PPh$_3$)$_3$] and [RuCl$_2$(PEt$_3$)$_3$].

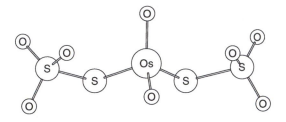

Fig. 4.13. The structure of [Os(O)$_2$(S$_2$O$_3$)$_2$]$^{2-}$, determined by X-ray diffraction for the [^{n}Bu$_4$N]$^+$ salt.

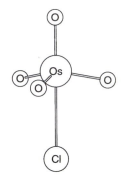

Fig. 4.14. The structure of [OsCl(O)$_4$]$^-$, determined by X-ray diffraction for the [Ph$_4$P]$^+$ salt.

Just as we have seen for earlier groups, the formulae of some simple halides may be misleading. Thus, OsF$_5$ and RuF$_5$ are tetramers with structure **4.1**, OsCl$_5$ is dimeric having structure **4.2**, and OsCl$_4$ is polymeric with chain structure **4.4**.

Coordination numbers above six are rare; OsF$_7$ has is pentagonal bipyramidal structure, and the complexes [OsH$_6$(PPh^{i}Pr$_2$)$_2$] and [OsH$_3$(PPh$_3$)$_4$] are two members of a series of hydrido species which feature high coordination numbers around osmium.

4.7 Rhodium and iridium

As seen in earlier groups, group 9 has examples where simple formulae are misleading in terms of structure. For example, RuF$_5$ is tetrameric and possesses structure **4.1**. Iridium pentafluoride was originally (pre-1965) formulated as IrF$_4$; it is tetrameric or polymeric.

Typical geometries for rhodium and iridium vary with oxidation state as shown in Table 4.2; square planar M(I) and octahedral M(III) complexes constitute particularly important families. The Rh(II) entry in Table 4.2 is an unusual example of stabilization of this oxidation state.

Lower oxidation states (−1, 0) are represented in organometallic complexes, and species for M(I) or M(II), e.g. *Vaska's compound* **4.5**, border between being coordination and organometallic in nature.

The structure of the rhodium(I) trigonal bipyramidal complex [Rh(SH){P(CH$_2$CH$_2$PPh$_2$)$_3$}] is shown in Fig. 4.15. This illustrates the use of a coordinatively restrictive *tripodal* ligand, i.e. a ligand bearing three arms (each with a donor-atom) which radiate from a central point, in this case, a central P atom.

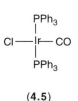

(4.5)

Table 4.2 Examples of group 9 metal coordination complexes: structures and oxidation states; common geometries are emphasized.

Oxidation state	Arrangement of donor atoms about metal centre	Examples for rhodium	Examples for iridium
+1	Square planar (*common*)	[RhCl(PPh$_3$)$_3$]	*trans*-[IrCl(CO)(PPh$_3$)$_2$]
	Trigonal bipyramidal	[Rh(SH){P(CH$_2$CH$_2$PPh$_2$)$_3$}]	[Ir(dppe)$_2$(CNMe)]$^+$
+2	Square planar	[RhCl$_2${P(*cyclo*-C$_6$H$_{11}$)$_3$}$_2$]	*trans*-[Ir(SEt)$_2$(Mes)$_2$]^a
+3	Five-coordinate	[RhH$_2$Cl(PPh$_3$)$_2$] (sbp)^b	[IrH$_3$(AsPh$_3$)$_2$] (tbp)^b
	Octahedral (*common*)	[RhCl$_6$]$^{3-}$; [Rh(CN)$_6$]$^{3-}$;	[IrCl$_6$]$^{3-}$; [Ir(CN)$_6$]$^{3-}$;
		[Rh(NH$_3$)$_6$]$^{3+}$; [Rh(N$_3$)$_6$]$^{3-}$;	[Ir(NH$_3$)$_6$]$^{3+}$; [Ir(H$_2$O)$_6$]$^{3+}$;
		[Rh(H$_2$O)$_6$]$^{3+}$; [Rh(ox)$_3$]$^{3-}$	*trans*-[IrCl$_2$(PMe$_2$Ph)$_4$]$^+$
+4	Octahedral	[RhF$_6$]$^{2-}$	[IrCl$_6$]$^{2-}$; [Ir(ox)$_3$]$^{2-}$
+5	Octahedral	[RhF$_6$]$^-$	[IrF$_6$]$^-$
+6	Octahedral	RhF$_6$	IrF$_6$

^{a}Mes = mesityl = 2,4,6-Me$_3$C$_2$; ^{b}sbp = square-based pyramidal; tbp = trigonal bipyramidal

Structure **4.6** shows the IrP$_4$ core of [Ir(PMePh$_2$)$_4$]$^+$; distortion away from a square planar environment is observed presumably caused by ligand crowding. This complex provides an excellent example of structures which deviate significantly from one of the classes listed in Table 4.2, and reminds us that these categories should not be regarded as restrictive choices for the metal centres. The cations [Rh(PMe$_2$Ph)$_4$]$^+$ and [Rh(PMe$_3$)$_4$]$^+$ are similarly distorted.

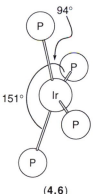

(4.6)

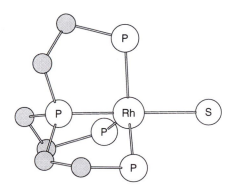

Fig. 4.15. The structure of [Rh(SH){P(CH$_2$CH$_2$PPh$_2$)$_3$}] determined by X-ray diffraction; the Ph groups and hydrogen atoms are omitted from the diagram.

4.8 Palladium and platinum

In comparison to what we have described for the heavier metals in groups 3-9, the group 10 metals palladium and platinum are restricted in the coordination geometries available. The coordination chemistry of Pd(II) and Pt(II) is almost invariably that of square-planar complexes while that of Pd(IV) and Pt(IV) is dominated by octahedral species. Platinum(VI) is rare, but is observed in the octahedral PtF_6, an extremely powerful oxidizing agent (e.g. it oxidizes O_2 to $[O_2]^+$).

The simplest examples of square planar M(II) complexes are halo complexes such as $[PdCl_4]^{2-}$ and $[PtCl_4]^{2-}$. With the exception of PdF_2, the neutral dihalides show square planar coordination, achieved by oligomerization. Palladium(II) chloride has two forms: α-$PdCl_2$ is polymeric (structure **4.7**) while β-$PdCl_2$ is hexameric (structure **4.8**); $PtCl_2$ behaves similarly. Palladium difluoride possesses a rutile lattice in which the Pd(II) centres are octahedrally sited, a rare situation for Pd(II).

Five-coordinate palladium(II) complexes are not common, but can be achieved if the ligand is able to control the coordination environment. Examples include the use of tripodal ligands (see Section 4.7) – cation **4.9** is one such complex. An instructive example of ligand-control involves the use of oligopyridine ligands. Figure 4.16 shows two ligands in this family; L^1 possesses four nitrogen donor atoms, while L^2 has five. The figure shows the conformations of the free ligands, but rotation about the transannular C–C bonds can occur, and does so upon complex formation. As expected, complexation between Pd(II) and L^1 produces a near square-planar cation (Fig. 4.17a). However, the availability of the fifth N atom in L^2 and the flexibility of the ligand backbone, encourages the formation of the 2:2 complex $[Pd_2(L^2)_2]^{4+}$ in which the two ligand chains adopt a double-helical arrangement (Fig. 4.17b). Five-coordinate geometries for platinum(II) are also rare.

4.9 Silver and gold

Whereas there are close similarities between palladium and platinum in their oxidation states and coordination geometries, the same is not true as we proceed to the next group. The coordination chemistry of silver is dominated by Ag(I); a much smaller range of Ag(II) complexes exists, and the chemistry of Ag(III) is very limited. Silver(I) shows flexibility in the geometries that it tolerates in its complexes, but low coordination numbers (two to four) are most common. However, octahedral sites are observed for Ag(I) in crystalline AgCl and AgBr which adopt NaCl-lattices, and octahedral coordination is also seen in some discrete complexes such as $[AgL_2]^+$ where ligand L is the macrocycle **4.10** (X = S). When silver(I) is four-coordinate, the geometry tends to be tetrahedral, e.g. in $[AgX(PPh_3)_3]$ (X = Cl, Br, I) and $[Ag(dppe)_2]^+$. Three-coordinate complexes are less common, and an example is $[AgTe_7]^{3-}$ which features a trigonal planar Ag(I) centre (Fig. 4.18).

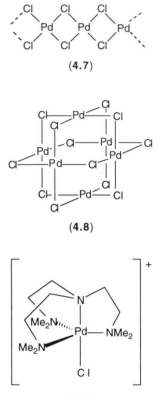

(4.7)

(4.8)

(4.9)

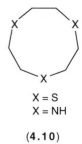

X = S
X = NH

(4.10)

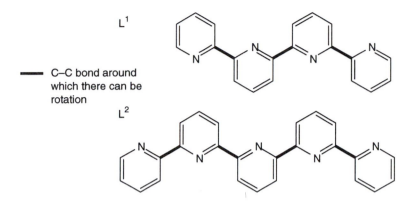

L^1

—— C–C bond around
which there can be
rotation

L^2

Fig. 4.16. Examples of oligopyridine ligands (see Fig. 4.17).

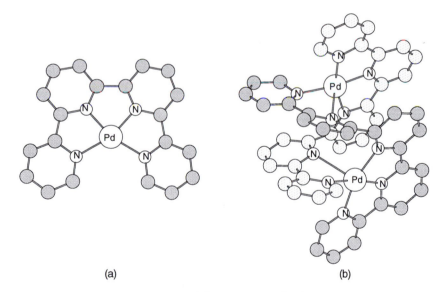

(a)   (b)

Fig. 4.17. The structures of the cations (a) [PdL^1]$^{2+}$ and (b) [Pd(L^2)$_2$]$^{4+}$ determined by X-ray diffraction for the hexafluorophosphate salts. The ligands are defined in Fig. 4.16. In (b), each ligand chain is distinguished by colour-coding to allow the double-helical nature of the complex to be apparent.

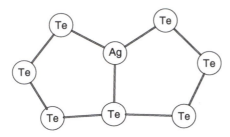

Fig. 4.18. The structure of $[AgTe_7]^{3-}$ determined by X-ray diffraction for the compound $[PPh_4]_2[NEt_4][AgTe_7]$.

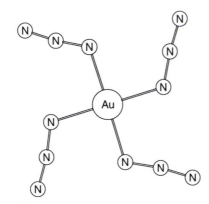

Fig. 4.19. The structure of $[Au(N_3)_4]^-$ determined by X-ray diffraction for the $[Ph_4As]^+$ salt.

In contrast to silver, gold forms a wide range of complexes in both the +1 and +3 oxidation states; the former typically exhibit linear coordination environments, while the latter are usually square planar. Examples for Au(I) are $[Au(CN)_2]^-$, $[R_3PAuCl]$, $[R_2SAuCl]$ and $[Au(PR_3)_2]^+$, and for Au(III) are $[AuCl_4]^-$, $[AuBr_4]^-$, $[Au(N_3)_4]^-$ (Fig. 4.19), $[(Et_3P)AuBr_3]$, $[Au(SeCN-Se)_4]^-$ and $[Au(en)_2]^+$. Less commonly, higher coordination numbers are observed.

The *relativistic contraction* of the $6s$ orbital in gold results in the distinctive properties of this metal compared to the lighter members of group 11. In particular, in multinuclear complexes, there is a tendency for the formation of Au–Au interactions. This manifests itself in a number of ways; Au–Au distances are variable, and may be taken in some cases to be 'bonds' and in others to be weaker interactions. It may also give rise to intermolecular Au⋯Au interactions in the solid state. Figure 4.20 shows an example of a trigold(I) complex in which the three metal centres interact with each other. This complex also gives an example of a three-coordinate Au(I) centre (ignoring the Au⋯Au contacts), a geometry which is well established but less common than linear Au(I) environments.

Relativistic effects: see page 52

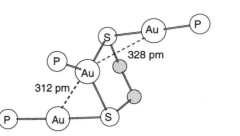

Fig. 4.20. The structure of [(Ph$_3$PAu)$_3$(SCH$_2$CH$_2$S)]$^+$ determined by X-ray diffraction for the tetrafluoroborate salt; Ph groups and H atoms have been omitted.

4.10 Cadmium and mercury

The presence of a filled *d*-shell for cadmium and mercury means that these metals are flexible in their coordination geometries; there is no ligand field stabilization effect. The +2 oxidation state is dominant for cadmium, while both the +1 and +2 states are important for mercury. However, mercury(I) is found in the form of the [Hg$_2$]$^{2+}$ ion in compounds such as Hg$_2$X$_2$ (X = F, Cl, Br, I), Hg$_2$SO$_4$ and Hg$_2$(NO$_3$)$_2$; the *coordination* chemistry of mercury is essentially that of Hg(II).

For cadmium(II), the most common coordination numbers are four, five and six. For example, [CdCl$_4$]$^{2-}$, [CdCl$_5$]$^{3-}$ and [CdCl$_6$]$^{4-}$ are tetrahedral, trigonal bipyramidal and octahedral, respectively. Six-coordinate complexes are very common, e.g. [Cd(H$_2$O)$_6$]$^{2+}$, [Cd(dmso-*O*)$_6$]$^{2+}$, [Cd(acac)$_3$]$^-$, *trans*-[Cd(en)$_2$(NCS-*S*)$_2$], [Cd(en)$_3$]$^{2+}$, [CdL$_2$]$^{2+}$ (L = **4.10** with X = NH) and [Cd(C$_5$H$_5$NO)$_6$]$^{2+}$. We have seen previous examples in this chapter of the donor atom arrangement being controlled by restrictions imposed by the ligand. The six-coordinate complex [Cd{H$_2$N(CH$_2$CH$_2$NH)$_4$CH$_2$CH$_2$NH$_2$}]$^{2+}$ is a further example; Fig. 4.21a illustrates severe distortion from an octahedral coordination environment.

Coordination numbers of seven and eight are known in a small number of cadmium(II) complexes. In some cases, as in [CdBr$_2$(18-crown-6)] (Fig. 14.21b), it is the demands of the ligand that result in the attainment of an unusual coordination number. The six *O*-donor atoms of the macrocyclic ligand are virtually coplanar with the Cd^{2+} ion also lying in the plane; the Br$^-$ ligands adopt axial sites to give a hexagonal bipyramidal structure.

Coordination numbers of two to six are the most usual for mercury(II). The metal centre is soft (i.e. class B) and many complexes involve sulfur donor ligands. Linear complexes include [Hg(NH$_3$)$_2$]$^{2+}$, [Hg(CN)$_2$] and [Hg{N(SO$_2$Me)$_2$}$_2$], while [HgI$_3$]$^-$ and [Hg(SPh)$_3$]$^-$ are examples of trigonal planar complexes. Four coordinate complexes may contain a tetrahedral Hg(II) centre as in [HgCl$_4$]$^{2-}$, [HgBr$_4$]$^{2-}$, [Cl$_2$Hg(μ-Cl)$_2$HgCl$_2$]$^{2-}$, [Hg(en)(SCN-*S*)$_2$] and [Hg(S$_6$)$_2$]$^{2-}$ (Fig. 4.22). However, chelation can lead to distorted square planar complexes, e.g. in bis(2-pyridineselenolato)mercury (Fig. 4.23) and bis(dithiocarbamate) complexes such as [Hg(S$_2$CNEt$_2$)$_2$] in which the   S-Hg-S angles are 66° (in the chelate ring) and 114°. Five-coordination is exemplified by [HgCl$_5$]$^{3-}$ (trigonal bipyramidal)   and [HgL]$^{2+}$ (Fig. 4.24) in

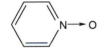

C$_5$H$_5$NO = pyridine-*N*-oxide

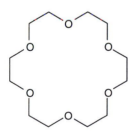

18-crown-6

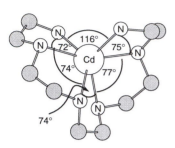

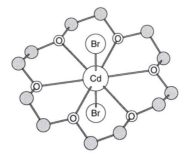

Fig. 4.21. The structures (determined by X-ray diffraction) of
(a) $[Cd\{H_2N(CH_2CH_2NH)_4CH_2CH_2NH_2\}]^{2+}$ (in the tetrafluoroborate salt) and
(b) $[CdBr_2(18\text{-crown-6})]$.

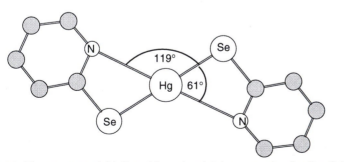

Fig. 4.22. The structure of the anion $[Hg(S_6)_2]^{2-}$ determined by X-ray diffraction for
the $[Et_4N]^+$ salt.

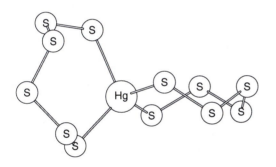

Fig. 4.23. The structure of bis(2-pyridineselenolato)mercury showing the distorted
square-planar environment caused by the chelation of each ligand.

which distortion is forced by the conformational restrictions of the macrocyclic ligand L. Six-coordinate complexes are usually octahedral, e.g. $[Hg(H_2O)_6]^{2+}$, $[Hg(py)_6]^{2+}$ and $[Hg(en)_3]^{2+}$.

Coordination numbers >6 are less common than those ≤6; a distorted square-antiprismatic coordination environment is seen in $[Hg(NO_2\text{-}O,O')_4]^{2-}$ As we have already seen, unusual coordination numbers may be forced by the nature of the ligands. For example, in $[Hg(H_2O)_2(NO_3)_2(py)_2]$, the close proximity of a second *O*-donor on each nitrato group results in a tendency towards eight-coordination (Fig. 4.25).

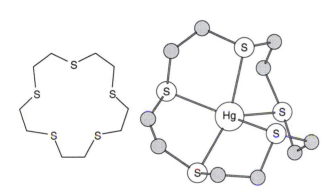

Fig. 4.24. The thia-crown ligand L forms a five-coordinate complex $[HgL]^{2+}$ which has a distorted coordination environment.

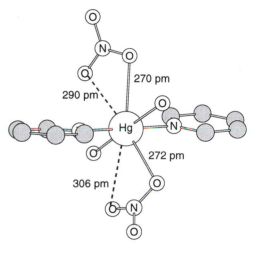

Fig. 4.25. The structure of $[Hg(H_2O)_2(NO_3)_2(py)_2]$ with H atoms omitted.

5 Coordination complexes: magnetic properties and electronic spectra

5.1 Introduction

This chapter deals with the role that the metal d-orbitals play in governing the properties of transition metal complexes. The chapter represents a mere introduction to the magnetic properties and electronic spectra of the heavier d-block metal coordination complexes; the topic is complicated and beyond the scope of this book, and so we focus on comparisons of properties between the metals of the first and later rows.

It is assumed that readers have already studied the chemistry of the first row d-block metals, and are already familiar with the concepts of crystal and ligand field theories, the splitting of the d-orbital set in transition metal complexes (Fig. 5.1) and the resulting effects on magnetic properties and electronic spectra. For complexes of the first row metals, octahedral and tetrahedral geometries tend to predominate. Octahedral complexes may be low- or high-spin, while tetrahedral complexes are high-spin because of the relatively small value of Δ_{tet}. A splitting diagram for square planar complexes is included in Fig. 5.1 because of its importance in the later metals of the d-block, particular for group 10 metal(II) ions (see below).

For an appropriate introduction with emphasis on first row d-block metals, see:
M. Gerloch and E.C. Constable (1994) *Transition Metal Chemistry: The Valence Shell in d-Block Chemistry*, VCH, Weinheim.

5.2 Metal nd-orbitals (n = 3, 4, 5)

Before considering the properties of metal complexes, we need to look at how the character of the metal d-orbitals changes in progressing from the first to third rows of the d-block. For the second and third row metals, the valence shell contains $4d$ and $5d$ atomic orbitals respectively. Compared to the $3d$ orbitals (which influence the chemistry of the *first* row metals), the $4d$ and $5d$ AOs are more diffuse and possess greater numbers of radial nodes. This is illustrated in Fig. 5.2 by schematic representations of the radial distribution functions for the $3d$ and $4d$ AOs. The probability of finding electrons *further away* from the nucleus increases in going from the $3d$ to $4d$ orbital. The consequence of these differences is that there is greater interaction between metal and ligand orbitals for heavier metals in a given triad of the d-block. Covalent interactions *between* metal centres also increase as we have already discussed in Section 2.5; we return to this in Chapter 6.

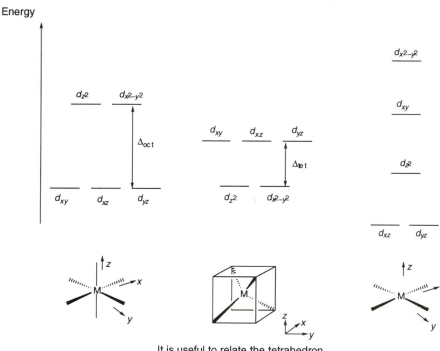

It is useful to relate the tetrahedron
to a cube in order to define the axis set clearly.

Fig. 5.1. A gaseous metal ion possesses a five-fold degenerate set of *d*-atomic orbitals; their energies are perturbed by the presence of ligands and the splitting of the *d*-orbital set depends on the spatial arrangement of the ligand donor atoms about the metal centre. Here we illustrate splitting for octahedral, tetrahedral and square planar arrangements.

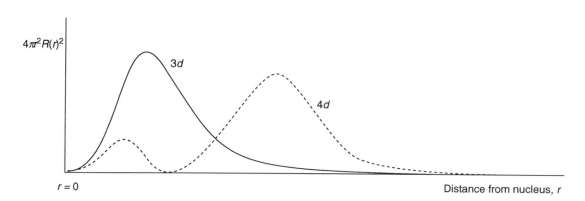

Fig. 5.2. Schematic representations of the radial distribution functions for the 3*d* and 4*d* atomic orbitals. What would the corresponding curve for the 5*d* radial distribution function look like?

5.3 Octahedral complexes

In order to gain some understanding of the differences in properties between first row metal complexes and those of their heavier congeners, we first take a look at some simple octahedral complexes containing metal ions from particular triads. Consider the complexes $[M(NH_3)_6]^{3+}$ and $[M(en)_3]^{3+}$ containing group 9 metals; values of Δ_{oct} are plotted in Fig. 5.3. The chemical relationship between ammine (NH_3) and diamine ($H_2N(CH_2)_nNH_2$) ligands would lead one to expect some similarity in the values of Δ_{oct} for $[M(NH_3)_6]^{3+}$ and $[M(en)_3]^{3+}$ for a *specified metal ion*. This is confirmed by the closeness of the points for pairs of complex ions in Fig. 5.3. The second feature of importance in the graph is that on moving from cobalt(III) to rhodium(III) (i.e. first row to second row) and from rhodium(III) to iridium(III) (i.e. second row to third row), there is a significant increase in the value of Δ_{oct}. This trend is repeated in other series of complexes. The consequence in terms of the occupancy of these orbitals in an octahedral complex is that low-spin complexes are energetically more favourable than high-spin complexes for second and third row metal ions. Let us look a bit more closely at this result, because for first row metal coordination complexes, much emphasis is placed on the *field strength of the ligands* in determining whether a ground state electronic configuration is high or low-spin.

The *spectrochemical series* is used to distinguish between strong field (e.g. CO) and weak field (e.g. I⁻) ligands:

$$I^- < Br^- < [SCN]^- < Cl^- < F^- < [OH]^- < [ox]^{2-} \approx H_2O < [NCS]^- < MeCN < NH_3 < [CN]^- < PR_3 < CO$$

Values of Δ are obtained from spectroscopic data, and are often quoted in units of cm^{-1}; the conversion to the SI unit of kJ mol^{-1} is:

$1\,cm^{-1} = 11.963 \times 10^{-3}$ kJ mol^{-1}

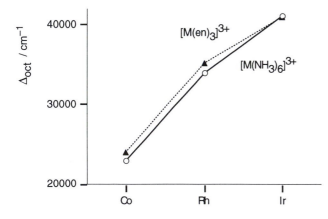

Fig. 5.3. Trends in values of Δ_{oct} for some related complexes of group 9 M(III) ions.

This series is useful in rationalizing observed differences in complexes of first row metal ions, for example, why $[Fe(CN)_6]^{4-}$ and $[Fe(CN)_6]^{3-}$ are low-spin, but $[FeF_6]^{3-}$ and $[FeCl_6]^{3-}$ are high-spin. However, when it comes to the ruthenium analogues of the halo anions, both $[RuF_6]^{3-}$ and $[RuCl_6]^{3-}$ are low-spin. This difference is a manifestation of the increased value of Δ_{oct} in going from iron(III) to ruthenium(III). Metal ions may also be arranged in a spectrochemical series:

$$Mn(II) < Ni(II) < Co(II) < Fe(III) < Cr(III) < Co(III) < Ru(III) < Mo(III) <$$
$$Rh(III) < Pd(II) < Ir(III) < Pt(IV)$$

and it is immediately apparent from this sequence that metal ions from the second and third rows correspond to the largest values of Δ_{oct}. Their influence is more dominant than that of the ligands, and as a result, octahedral metal complexes containing the heavier metal ions are invariably low-spin.

5.4 Tetrahedral and square planar complexes

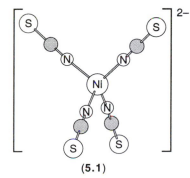

(5.1)

For complexes of first row metal ions having a d^8 configuration (e.g. Ni^{2+}), the preference of tetrahedral over square planar (or vica versa) is affected by the types of ligands present. For example, $[Ni(NCS\text{-}N)_4]^{2-}$ (**5.1**), $[Ni(NCS\text{-}S)_4]^{2-}$, $[NiCl_4]^{2-}$, $[NiBr_4]^{2-}$ and $[NiI_4]^{2-}$ are tetrahedral and paramagnetic (Fig. 5.4a), whereas $[Ni(CN)_4]^{2-}$, **5.2**, (containing strong field ligands) is square planar and diamagnetic (Fig. 5.4b). On moving down group 10, it is found that $[Pd(CN)_4]^{2-}$, $[Pt(CN)_4]^{2-}$, $[PdBr_4]^{2-}$, $[PtBr_4]^{2-}$, $[PdCl_4]^{2-}$ and $[PtCl_4]^{2-}$ are *all* square planar, as, indeed, are the majority of Pd(II) and Pt(II) complexes.

(5.2)

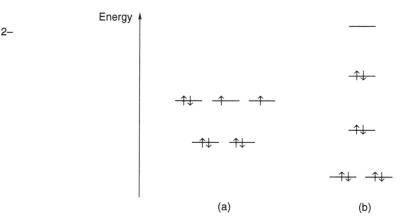

Fig. 5.4. A d^8 metal complex is (a) paramagnetic if it is tetrahedral, and (b) diamagnetic if it is square planar; refer also to Fig. 5.1.

It is worth commenting upon the complexation of Ni(II), Pd(II) and Pt(II) with, for example, ammonia. Whereas nickel(II) forms octahedral $[Ni(NH_3)_6]^{2+}$, the heavier metals form the square planar $[Pd(NH_3)_4]^{2+}$ and $[Pt(NH_3)_4]^{2+}$ cations.

5.5 Limitations of the spin-only formula

The effective magnetic moment of an octahedral complex of a first row d-block metal ion can be estimated using the *spin-only formula*, two forms of which are given in Eqns 5.1 and 5.2.

$$\text{Spin - only magnetic moment} = \mu(\text{spin - only}) = \sqrt{n(n+2)} \tag{5.1}$$

where n = number of unpaired electrons

$$\text{Spin - only magnetic moment} = \mu(\text{spin - only}) = \sqrt{4S(S+1)} \tag{5.2}$$

where S = total spin quantum number

The formula is only approximate, but gives good estimates for octahedral complexes of first row metal ions, e.g. for $[Mn(CN)_6]^{2-}$, the experimentally observed μ_{eff} = 3.94 BM and $\mu(\text{spin-only})$ = 3.87 BM (Eqn 5.3).

$$\mu(\text{spin-only}) = \sqrt{3(3+2)} = \sqrt{15} = 3.87 \tag{5.3}$$

Values of $\mu(\text{spin-only})$ for complexes with first row M^{n+} ions in environments other than octahedral often show significant deviations from experimental values of μ_{eff}. For example, for tetrahedral $[NiCl_4]^{2-}$, μ_{eff} = 3.89 BM compared to $\mu(\text{spin-only})$ = 2.83 BM. For complexes of second and third row metal ions, irrespective of the arrangement of the donor atoms around the metal centre, using the spin-only formula produces values of $\mu(\text{spin-only})$ which do not correspond to experimental values of μ_{eff}. The reason for the breakdown is that equations 5.1 and 5.2 can only be relied upon when the magnitude of the magnetic moment is governed solely or predominantly by *spin angular momentum*. In practice, it is the total angular momentum which determines the magnetic susceptibility, i.e. contributions both from spin *and* orbital angular momenta. In many octahedral complexes of first row metal ions, the orbital angular momentum is *quenched*.

The quantum numbers J, S and L describe the total, spin and orbital angular momenta respectively. Both Eqns 5.4 and 5.5 (the van Vleck formula) are also used to estimate magnetic moments.

Magnetic susceptibility, χ, and effective magnetic moment, μ_{eff} are related by the equation:

$$\mu_{\text{eff}} = 2.828\sqrt{\chi T}$$

where T = temperature in kelvin.

$$\mu_{eff} = g\sqrt{J(J+1)}$$

(5.4)

where: $g = 1 + \dfrac{J(J+1) + S(S+1) + L(L+1)}{2J(J+1)}$

$$\mu_{eff} = \sqrt{L(L+1) + 4S(S+1)}$$

(5.5)

In Eqn 5.4, the effects of *spin-orbit coupling* (i.e. coupling between spin and orbital angular momenta) are taken into account, while in Eqn 5.5, they are ignored. While these equations are an improvement upon the spin-only formula, and Eqn 5.4 may be applied with success to lanthanoid ions, neither equation gives good estimates of magnetic moments for second and third row metal ions. Spin-orbit coupling increases significantly down a triad in the *d*-block, complicating the magnetic behaviour.

Table 5.1 lists room temperature magnetic data for selected octahedral complexes of Ru(III), Nb(IV) and Ta(IV), all of which possess one unpaired electron; low-spin Ru(III) has the configuration t_{2g}^5, low-spin Nb(IV) and Ta(IV) are t_{2g}^1. Two points should be noted:
- the relatively wide range of observed values;
- the significant deviation from the calculated spin-only values.

Indeed, if notice were being paid to the calculated values to attempt to say something about metal oxidation state, we would be in trouble! Similar problems arise throughout the second and third row. A further example is seen with the low-spin Re(III) and Re(V) complexes *cis*-[ReCl$_4$(NCMe)$_2$]$^-$ and [Re(O)Cl$_5$]$^{2-}$ which have electronic configurations of t_{2g}^4 and t_{2g}^2

Table 5.1 Examples of the breakdown of the spin-only formula for selected octahedral complexes of the heavier *d*-block metal ions.

Complex	Experimental μ_{eff} / BM (Temp / K)	μ(spin-only) / BM
[Ru(NH$_3$)$_6$]$^{3+}$	2.14 (300)	1.73
[RuCl$_6$]$^{3-}$	2.24 (298)	1.73
[NbF$_6$]$^{2-}$	1.30 (293)	1.73
cis-[NbBr$_4$(NCMe)$_2$]	1.27 (300)	1.73
cis-[TaCl$_4$(NCMe)$_2$]	0.45 (293)	1.73

respectively, i.e. both with two unpaired electrons. The observed values (room temperature) of μ_{eff} are 1.70 and 0.52 BM respectively; using the spin-only formula gives an unrealistic value of 2.83 BM.

A number of the examples above have shown experimental room temperature values of μ_{eff} which are much smaller than spin-only values. On their own, these data are, effectively, meaningless, and the measurement of magnetic moments over a wide range of temperatures is essential for second and third row metal ions. The value of μ_{eff} is highly dependent on temperature when the spin-orbit coupling constant is large — a characteristic of the heavier metals ions. In order to illustrate this, we take a series of ions with t_{2g}^4 configurations and consider a plot (a so-called *Kotani plot*, Fig. 5.5) of μ_{eff} against $(-kT / \lambda)$ where k is the Boltzmann constant, T is temperature, and λ is the spin-orbit coupling constant (a negative number). For Cr(II) and Mn(III), λ is small, but is much larger for Ru(IV) and is larger still for Os(IV). For a given temperature, kT is a constant, the value of μ_{eff} decreases as the spin-orbit coupling gets larger. The points indicated on the plot in Fig. 5.5 mark the room temperature values of μ_{eff} for the t_{2g}^4 ions. Notice that the first row metal ions lie on an essentially horizontal part of the curve, i.e. changes in temperature have little effect on the value of μ_{eff}. In contrast, the second and third row ions lie on parts of the curve with steep gradients and their magnetic moments will be greatly affected by temperature changes. Osmium(IV) is a particularly dramatic example of the temperature dependence of the magnetic moment.

Boltzmann constant,
$$k = 1.38 \times 10^{-23} \text{ J K}^{-1}$$

Discussions of magnetism cannot be covered in depth in this book. Suggested references for more detailed treatments are:
S.F.A. Kettle (1996) *Physical Inorganic Chemistry*, Spektrum, Oxford — Chapter 9 deals with magnetic properties of *d*-block metals.
M. Gerloch (1986) *Orbitals, Terms and States*, Wiley, Chichester — Chapter 4 deals in detail with spin-orbit coupling.

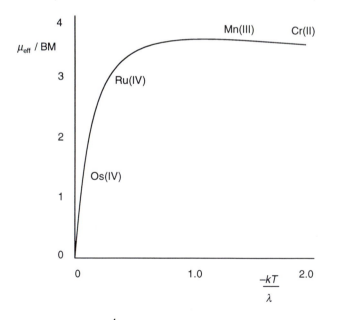

Fig. 5.5. Kotani plot for a t_{2g}^4 configuration; the points marked show 298 K data for selected ions. Notice that even the first row ions do not obey the spin only formula and that orbital contributions to the magnetic moment are present.

5.6 Electronic spectra

From studies of first row *d*-block metal complexes, readers will be familiar with the Laporte forbidden '*d-d*' transitions, and the fact that these give rise to *weak* absorptions in electronic spectra. The second type of electronic transition of importance involves the transition from a metal-centred to ligand-centred molecular orbital, or *vice versa*. These are *charge transfer* transitions, and give rise to much more intense absorptions in the electronic spectrum of a complex than '*d-d*' transitions. Intensity of absorption is described by the extinction coefficient, ε_{max} (Fig. 5.6), which is generally $>10^3$ dm^3 mol^{-1} cm^3 for charge transfer bands. A well known example for a first row metal is the deep purple colour of KMnO$_4$ solutions, caused by a ligand-to-metal charge transfer (LMCT) transition. For complexes of the heavier metal ions, CT absorptions are often the most important features of the electronic spectra.

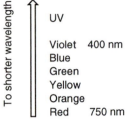

Some ligands also absorb in the UV: e.g. ox^{2-} (250 nm), [N$_3$]$^-$ (235 nm), I$^-$ (226 nm), [SCN]$^-$ (213 nm)

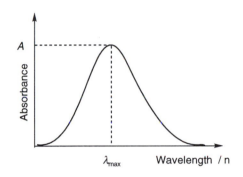

Fig.5.6. An absorption band in an electronic spectrum is characterized by the wavelength *and* absorbance of the band maximum. The value of *A* marked on the absorbance axis can be converted to the correponding value of ε_{max} by using the Beer-Lambert Law.

LMCT = metal-to-ligand charge transfer
MLCT = ligand-to-metal charge transfer

Figure 5.7 shows an approximate MO diagram for an octahedral [ML$_6$]$^{n+}$ complex; the occupancies of the MOs depend on the electronic configuration of the metal ion. The MOs may be characterized as being metal- or ligand-centred since their composition will depend upon the relative energies of the metal (left hand side of Fig. 5.7) or ligand (right hand side of Fig. 5.7) orbitals. If electrons occupy the metal-centred t_{2g} orbitals and ligand-centred orbitals such as the band (e.g. σ^*) shown are unoccupied, then an MLCT transition may occur. The wavelength of an MLCT or LMCT band (between which it is often hard to distinguish) is sensitive to the difference in electronegativity between the ligand and metal centres, since this affects the energies of the metal and ligand orbitals and, in turn, the energies of the MOs involved in the electronic transition. This is best illustrated by considering several specific series of complexes.

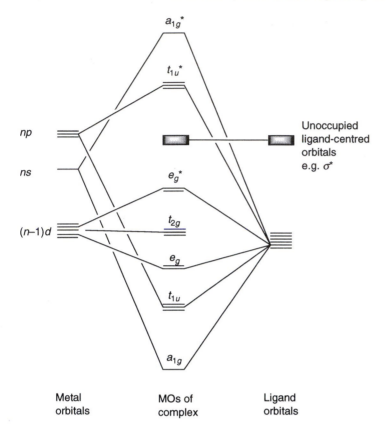

np

ns

(n–1)d

a_{1g}^*

t_{1u}^*

Unoccupied
ligand-centred
orbitals
e.g. σ^*

e_g^*

t_{2g}

e_g

t_{1u}

a_{1g}

Metal
orbitals

MOs of
complex

Ligand
orbitals

Fig.5.7. Schematic MO diagram for the formation of an octahedral $[ML_6]^{n+}$ complex.

Each of the hexahalo complexes $[OsCl_6]^{2-}$, $[OsBr_6]^{2-}$ and $[OsI_6]^{2-}$ contains an osmium(IV) centre. The decrease in electronegativity of the halo ligand (Cl > Br > I) results in a shift to longer wavelength for the CT band. This follows from the fact that an increase in electronegativity of the ligand lowers the energy of the ligand orbitals, thereby increasing the energies of charge transfer transitions:

$$\Delta E(Cl) > \Delta E(Br) > \Delta E(I)$$

Since E is inversely proportional to the wavelength, λ, the wavelength of the CT band follows the sequence:

$$\lambda(Cl) < \lambda(Br) < \lambda(I)$$

While attempts may be made to make other correlations such as changes in colour of related complexes as a group of metals is descended (e.g. purple $[MnO_4]^-$ to colourless $[TcO_4]^-$ and colourless $[ReO_4]^-$), much care is needed. For example, the effects that the counterion may have on the colour of a salt must be considered.

$E = h\nu$

$c = \lambda\nu$

Caution! Watch the units for wavelength; this sequence is true for λ in the correct units (e.g. nm) but is reversed if wavenumbers (cm^{-1}) are used.

6 Multiply bonded dinuclear complexes

6.1 Metal-metal bonding: general comments

For a detailed review on relativistic effects, see:
P. Pyykkö (1988) *Chemical Reviews*, **88**, 563.

In Section 2.5, we discussed trends in standard enthalpies of atomization and metal-metal bonding among the *d*-block metals. In descending a triad of metals, metal-metal bonding becomes increasingly important. In main group chemistry, an expected trend is for bond energies to *decrease* down a group of *p*-block elements. However, within the *d*-block, M–M bond energies have a tendency to *increase* down a triad. This observation can be rationalized in terms of *relativistic effects*. This is far from a simple concept, and we shall merely state that the effects of combining Einstein's theory of relativity with quantum theory result in an understanding of why *d*-orbitals are subject to a *relativistic expansion* which is more significant for 5*d* than for 4*d* than for 3*d* AOs.

6.2 Metal-metal quadruple and triple bonds

For the original report of the quadruple bond in $[Re_2Cl_8]^{2-}$, see:
F.A. Cotton *et al.* (1964) *Science*, vol. 145, p. 1305.

The term 'multiple bond' is most often associated with double and triple bonds. However, it is now well established that the overlap of appropriate *d*-orbitals can lead to the formation of metal-metal quadruple bonds. The first example to be reported was $[Re_2Cl_8]^{2-}$, **6.1**, and a key structural feature which pointed towards quadruple bond formation was the sterically unfavourable *eclipsed* configuration of the chloro ligands (Fig. 6.1).

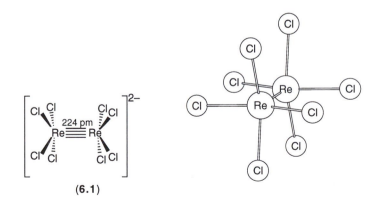

(6.1)

Fig. 6.1. The structure of $[Re_2Cl_8]^{2-}$ shown schematically and from crystallographic data emphasizing the eclipsed configuration of the chloro ligands.

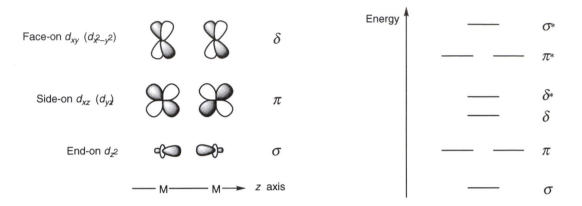

Fig. 6.2. Overlap of metal d-orbitals leading to the formation of σ, π and δ interactions.

Let us take $[Re_2Cl_8]^{2-}$ as an example, and define the Re atoms to lie on the z axis. Each Re centre uses four of its nine AOs (s, p_x, p_y and $d_{x^2-y^2}$) to form Re–Cl bonds. We may then allow mixing of the p_z and d_{z^2} orbitals to give two hybrid orbitals which point along the z axis. Each Re centre has four orbitals available for metal-metal bonding: d_{xz}, d_{yz}, d_{xy} and one p_z/d_{z^2} hybrid; the second hybrid remains non-bonding in $[Re_2Cl_8]^{2-}$ and points outwards from the M–M unit (see structure **6.6**). Overlap of the metal pd hybrid orbitals leads to the formation of a σ bond; d_{xz}–d_{xz} and d_{yz}–d_{yz} orbital overlap gives rise to a degenerate pair of π molecular orbitals (Fig. 6.2). The remaining AO is the d_{xy} (see margin), and 'face-on' interaction between these orbitals results in the formation of a δ bond (Fig. 6.2). The degree of overlap follows the order:

$$\sigma > \pi > \delta$$

and the approximate relative energies of the σ, π, δ, σ^*, π^* and δ^* MOs are shown schematically in Fig. 6.2. The pattern of occupancy of these orbitals is now crucial to the bond order. An electronic configuration $\sigma^2\pi^4\delta^2$ leads to a quadruple bond, but too few or too many electrons results in the loss of the δ component. In $[Re_2Cl_8]^{2-}$, the σ, π, δ levels (Fig. 6.2) are fully occupied and the effect is two-fold:
  • a rhenium-rhenium quadruple bond is present;
  • the δ component of the bond forces the two $ReCl_4$ units to be eclipsed.
The $[Tc_2Cl_8]^{2-}$ ion similarly exhibits an eclipsed configuration with a short (215 pm) Tc–Tc bond length.
    A point that needs emphasizing is that the diamagnetic character of salts containing $[Re_2Cl_8]^{2-}$ (or related anions) is consistent with the bonding picture described above. Further, salts of $[Re_2Cl_8]^{2-}$ are blue ($\lambda_{max} = 700$ nm) and this can be rationalized in terms of the δ–δ^* energy gap and a $\sigma^2\pi^4\delta^1\delta^{*1} \leftarrow \sigma^2\pi^4\delta^2$ transition.

The 'choice' between use of d_{xy} or $d_{x^2-y^2}$ for Re–Cl bonding is a consequence of the axis definition in respect of the Cl positions.

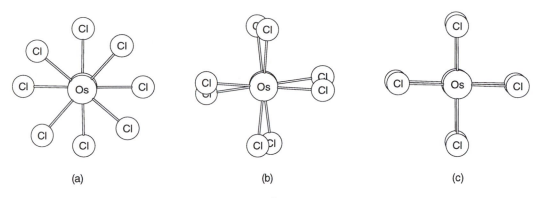

(a)   (b)   (c)

Fig. 6.3. The arrangement of the chloro ligands in $[Os_2Cl_8]^{2-}$ is not fixed. Solid state structure determinations show that, for example, (a) a staggered configuration is observed in the $[Bu_4N]^+$ salt, (b) the configuration in the $[(Ph_3P)_2N]^+$ salt is near to being eclipsed, and (c) the chloro ligands are fully eclipsed in the $[MePh_3P]^+$ salt. All structures are viewed approximately along the metal-metal axis.

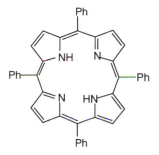

meso-tetraphenylporphyrin

Although we have drawn attention to the eclipsing of ligands being a consequence of quadruple bond formation, studies with porphyrin complexes have provided data information regarding rotation about this bond. The complex $[Mo_2(TPP)_2]$ (H_2TTP = *meso*-tetraphenylporphyrin) contains two {Mo(TTP)} units connected by a Mo–Mo quadruple bond, and from variable temperature NMR spectroscopic data, it has been estimated that the activation energy for rotation about the bond is ≈ 45 kJ mol^{-1}. Interestingly, in the solid state, the two $[TTP]^{2-}$ ligands are mutually twisted by 18° and the Mo–Mo quadruple bond is unusually long (224 pm; compare values in Table 6.1).

The addition of two electrons to the system can effectively be achieved by changing the metal from Re (or Tc) to Os. In $[Os_2Cl_8]^{2-}$, the δ^* level is fully occupied giving a configuration $\sigma^2\pi^4\delta^2\delta^{*2}$. In accordance with the 'cancelling out' of the δ bond by the occupancy of the corresponding antibonding MO and, thus, the presence of a metal-metal *triple* bond , Figure 6.3 shows that there is no restriction on the configuration that the chloro ligands can adopt. A similar situation is observed for $[Os_2Br_8]^{2-}$, but for $[Os_2I_8]^{2-}$ which contains more bulky substituents, only the staggered configuration has so far been observed in the solid state. For the group 6 metals, the representative members of this family of halo compounds are $[Mo_2Cl_8]^{4-}$ and $[W_2Cl_8]^{4-}$ which have the same number of electrons available for metal-metal bonding as $[Re_2Cl_8]^{2-}$. Thus, we again see the formation of a quadruple bond (Mo–Mo = 214 pm; W–W = 226 pm) and the eclipsing of ligands.

The synthesis of $[Re_2Cl_8]^{2-}$ involves H_3PO_2 reduction of $[ReO_4]^-$ in aqueous HCl, and halogen exchange may be used to obtain $[Re_2Br_8]^{2-}$. The $[Os_2Cl_8]^{2-}$ and $[Os_2Br_8]^{2-}$ ions can be prepared from $[Os_2(O_2CMe)_4Cl_2]$ and HCl or HBr respectively under rigorously anhydrous conditions. The triply bonded complex $[Os_2(O_2CMe)_4Cl_2]$, **6.2** (prepared by treating $[OsCl_6]^{2-}$ with $MeCO_2H$ and $(MeCO)_2O$) is a useful precursor for a range of diosmium species and introduces a general class of compound in which the multiply bonded dimetal unit is supported by bridging ligands. A synthetically important member of this family is $[Mo_2(O_2CMe)_4]$, **6.3**, which formally contains the quadruply bonded $[Mo_2]^{4+}$ unit. Compound **6.3** is prepared from $[Mo(CO)_6]$ and $MeCO_2H$ (or $MeCO_2H$ and $(MeCO)_2O$), and treatment of **6.3** with KCl in HCl yields $[Mo_2Cl_8]^{4-}$. Ligand replacement according to reactions 6.1 and 6.2 allows the introduction of $[SO_4]^{2-}$ or $[HPO_4]^{2-}$ bridges.

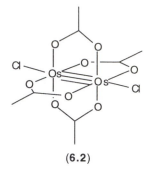

(6.2)

$$[Mo_2Cl_8]^{4-} + 4[SO_4]^{2-} \rightarrow [Mo_2(SO_4)_4]^{4-} + 8Cl^- \qquad (6.1)$$

$$[Mo_2Cl_8]^{4-} + 4[HPO_4]^{2-} \xrightarrow{\text{In presence of } O_2} [Mo_2(HPO_4)_4]^{2-} + 8Cl^- \quad (6.2)$$

(6.3)

The series of ions $[Mo_2Cl_8]^{4-}$, $[Mo_2(SO_4)_4]^{4-}$ and $[Mo_2(HPO_4)_4]^{2-}$ is instructive in terms of metal-metal bonding. We have already seen that $[Mo_2Cl_8]^{4-}$ possesses a quadruple bond by virtue of a $\sigma^2\pi^4\delta^2$ configuration, and $[Mo_2(SO_4)_4]^{4-}$ is analogous. On the other hand, reaction 6.2 involves aerial oxidation and $[Mo_2(HPO_4)_4]^{2-}$ has the electronic configuration $\sigma^2\pi^4$ and, thus, contains a metal-metal triple bond. An intermediate state is

Table 6.1 Metal-metal bond lengths and orders in selected dimolybdenum species.

Species[a]	Mo–Mo distance / pm	Mo–Mo bond order	Comments
$[Mo_2(O_2CMe)_4]$	209	4.0	
$[Mo_2(O_2CCF_3)_4]$	209	4.0	
$[Mo_2Cl_8]^{4-}$	214	4.0	
$[Mo_2(SO_4)_4]^{4-}$	211	4.0	
$[Mo_2(SO_4)_4(H_2O)_2]^{3-}$	217	3.5	Contains axial H_2O ligands
$[Mo_2(HPO_4)_4(H_2O)_2]^{2-}$	223	3.0	Contains axial H_2O ligands

[a]Anions crystallographically characterized as K^+ salts.

produced if $[Mo_2(SO_4)_4]^{4-}$ is oxidized by O_2 to $[Mo_2(SO_4)_4]^{3-}$; the loss of one electron gives a configuration of $\sigma^2\pi^4\delta^1$ and a bond order of 3.5. These data are summarized in Table 6.1 along with metal-metal bond lengths derived from X-ray diffraction studies; the incorporation of axial ligands follows from lone pair donation into the unused, outward pointing metal hybrid orbital that we allocated in our earlier discussion. Similar results are observed in related systems; e.g. oxidation of $[Mo_2(O_2CMe)_4]$ with I_2 generates the paramagnetic $[Mo_2(O_2CMe)_4]^+$ in which the Mo–Mo bond order is 3.5 ($\sigma^2\pi^4\delta^1$ configuration).

For a detailed discussion, see: F. A. Cotton and R. A. Walton (1993) *Multiple Bonds between Metal Atoms*, 2nd Edn, Oxford University Press, Oxford.

The number of compounds containing M–M triple or quadruple bonds has increased dramatically over the past two decades, with molybdenum and tungsten being key metals. We consider some aspects of the chemistry of selected compounds in the remaining part of the chapter. However, before leaving this section, we must draw attention to a family of dirhodium complexes of the types shown in **6.4** and **6.5**.

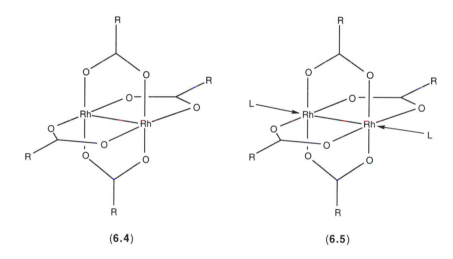

(6.4)   (6.5)

Although these resemble the carboxylate bridged dimers mentioned above, the dirhodium compounds contain rhodium-rhodium *single* bonds. Equation 6.3 shows a general preparative method, and the incorporation of axial ligands has the potential for the synthesis of polymeric materials such as the two-dimensional polymer $[\{Rh_2(O_2CCF_3)_4\}.TCNE]_n$ (Fig. 6.4).

TCNE = tetracyanoethene
$(NC)_2C=C(CN)_2$

$$RhCl_3 \cdot 3H_2O \xrightarrow{\text{RCO}_2\text{H, RCO}_2\text{Na}} Rh_2(O_2CR)_4 \qquad (6.3)$$

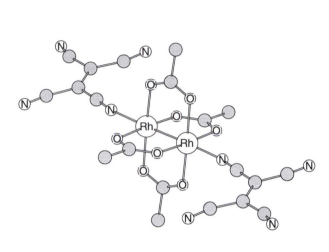

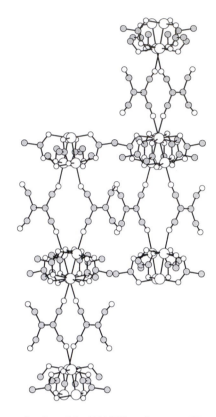

Fig. 6.4. The solid state structure of [{Rh$_2$(O$_2$CCF$_3$)$_4$}. TCNE]$_n$: the axial coordination of the TCNE ligands to one Rh$_2$ unit, and part of the packing diagram showing the generation of a polymer. The F atoms of the CF$_3$ groups have been omitted for clarity.

6.3 [Mo$_2$(O$_2$CR)$_4$]

The compound [Mo$_2$(O$_2$CMe)$_4$], **6.3**, is one of a group of quadruply bonded dimolybdenum species which contain four carboxylate bridges. It is a valuable precursor to many compounds with dimolybdenum cores. Reaction types can be summarized as:

- addition of axial ligands (adduct formation);
- replacement of the acetate ligands;
- reactions involving oxidation of the [Mo$_2$]$^{4+}$ core (see above);
- reactions which cleave the metal-metal bond.

R

O ⎓ O
−

Carboxylate
e.g. R = Me, CF$_3$, Et, ^{t}Bu, Ph

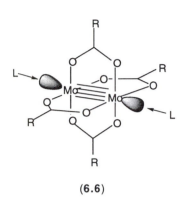

(6.6)

The addition of axial ligands arises because of the available vacant orbital per metal as shown in structure **6.6**. We have already seen (Table 6.1) the presence of axial H_2O ligands in $[Mo_2(SO_4)_4(H_2O)_2]^{3-}$ and $[Mo_2(HPO_4)_4(H_2O)_2]^{2-}$, but the formation of '$[Mo_2(O_2CMe)_4(H_2O)_2]$' has not been observed; indeed, adding axial ligands to the $\{Mo_2\}^{4+}$ core turns out to be rather difficult. An unstable adduct $[Mo_2(O_2CMe)_4(py)_2]$ is formed between $[Mo_2(O_2CMe)_4]$ and pyridine, and a more stable one between pyridine and $[Mo_2(O_2CCF_3)_4]$. The solid state structure of $[Mo_2(O_2CCF_3)_4(py)_2]$ reveals long Mo–N bonds (255 pm) and a lengthening (see Table 6.1) of the Mo–Mo bond to 213 pm. Such significant lengthening is not necessarily observed. An interesting example of axial ligand coordination is seen in the solid state structure of $K_4[Mo_2(SO_4)_4]\cdot2H_2O$ where terminal oxygen atoms from the sulfate bridges of one anion enter into axial coordination interactions with Mo centres on adjacent anions (Fig. 6.5).

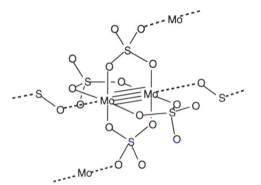

Fig. 6.5. Schematic illustration of inter-anion O–Mo axial interactions which generates a network in the solid state structure of $K_4[Mo_2(SO_4)_4]\cdot2H_2O$.

Morpholine

The replacement of acetate bridges is synthetically useful as a means of preparing a range of derivatives; replacement can be total or partial as the examples in Fig. 6.6 illustrate. While reaction of $[Mo_2(O_2CMe)_4]$ with KCl in HCl yields $[Mo_2Cl_8]^{4-}$, treatment with morpholine hydrochloride in concentrated HCl yields the morpholinium salt of $[Mo_2Cl_6(H_2O)_2]^{2-}$ which has the configuration shown in **6.7**. The bromo complex $[Mo_2Br_6(H_2O)_2]^{2-}$ (Mo–Mo = 213 pm) can be similarly prepared. The introduction of $[MeSO_3]^-$ or $[CF_3SO_3]^-$ bridging ligands gives synthetically useful anions, and reaction 6.4 shows the formation of the highly reactive $[Mo_2(NCMe)_8]^{4+}$ cation, which may be isolated as the blue salts $[Mo_2(NCMe)_8][CF_3SO_3]_4$ or $[Mo_2(NCMe)_8][BF_4]_4$.

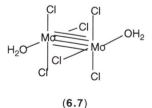

(6.7)

$$[Mo_2(H_2O)_4(O_3SCF_3)_2]^{2+} \xrightarrow{\text{MeCN}} [Mo_2(NCMe)_8]^{4+} \qquad (6.4)$$

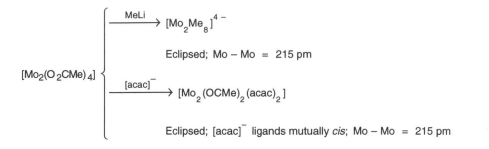

Fig. 6.6. Two reactions of [Mo$_2$(O$_2$CMe)$_4$] which illustrate total and partial replacement of carboxylate ligands.

So far, we have only mentioned reactions in which the dimolybdenum core is retained, but cleavage of this unit occurs with a number of reagents. For example, treatment of [Mo$_2$(O$_2$CMe)$_4$] with MeNC affords seven-coordinate [Mo(CNMe)$_7$]$^{2+}$ which has a capped octahedral structure.

6.4 [Re$_2$Cl$_8$]$^{2-}$

The chemistry of [Re$_2$Cl$_8$]$^{2-}$ (Fig. 6.1) is well developed and in this section we give examples that illustrate aspects of ligand displacement and redox reactions. The reaction between Cl$_2$ and [Re$_2$Cl$_8$]$^{2-}$ in MeCN proceeds with oxidation and chloride addition to give [Re$_2$Cl$_9$]$^-$. The Re–Re distance is 270 pm and three chloro bridges support the dimetal framework as shown in Fig. 6.7.

The reactions between [Re$_2$Cl$_8$]$^{2-}$ and carboxylates give compounds of type [Re$_2$(O$_2$CR)$_4$Cl$_2$] in which the chloro ligands occupy axial sites and a rhenium-rhenium quadruple bond is retained (compare structure **6.2**, and consider the periodic relationship between Re and Os). Treating [Re$_2$(O$_2$CR)$_4$Cl$_2$] with HCl regenerates [Re$_2$Cl$_8$]$^{2-}$.

Reactions with phosphines proceed with concomitant reduction:

$$[Re_2Cl_8]^{2-} \rightarrow [Re_2Cl_4(PR_3)_4]$$

formally:

$$\{Re_2\}^{6+} + 2e^- \rightarrow \{Re_2\}^{4+}$$

$$\text{i.e. } \sigma^2\pi^4\delta^2 \rightarrow \sigma^2\pi^4\delta^2\delta^{*2}$$

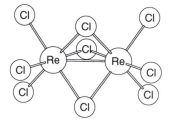

Fig. 6.7. The structure of [Re$_2$Cl$_9$]$^-$ determined by X-ray diffraction for the tetrabutylammonium salt.

and a change in the Re–Re bond order from 4.0 to 3.0. Although the metal-metal bond length in [Re$_2$Cl$_4$(dppe)$_2$] is the same as in [Re$_2$Cl$_8$]$^{2-}$ (224 pm), the arrangement of the ligands (Fig. 6.8) is consistent with the presence of a triple bond. [Re$_2$Cl$_4$(dppe)$_2$] can be made either by replacement of PEt$_3$ in [Re$_2$Cl$_4$(PEt$_3$)$_4$] by dppe, or by direct reaction of [Re$_2$Cl$_8$]$^{2-}$ with dppe. In terms of 'clamping' two metal centres together, it is well recognized that

Ligand abbreviations: see page 87.

the dppm ligand has a particular stabilizing effect. A number of dirhenium compounds containing dppm have been structurally characterized and these include those in the series $[Re_2Cl_4(dppm)_2]$, $[Re_2Cl_5(dppm)_2]$ and $[Re_2Cl_6(dppm)_2]$ (Fig. 6.8). Along this series, the Re–Re bond lengthens, consistent with a weakening of the bond.

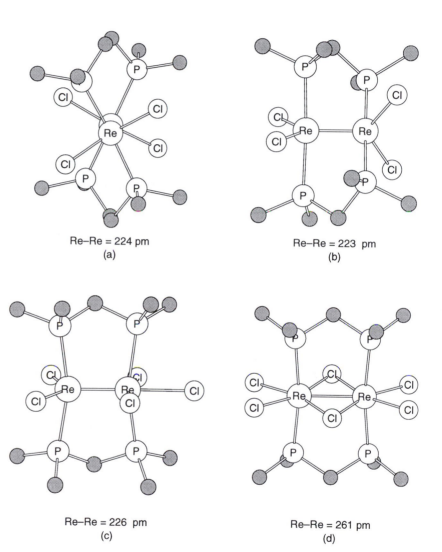

Re–Re = 224 pm
(a)

Re–Re = 223 pm
(b)

Re–Re = 226 pm
(c)

Re–Re = 261 pm
(d)

Fig. 6.8. The structures (determined by X-ray crystallography) of
(a) $[Re_2Cl_4(dppe)_2]$, (b) $[Re_2Cl_4(dppm)_2]$, (c) $[Re_2Cl_5(dppm)_2]$ and (d) $[Re_2Cl_6(dppm)_2]$;
for clarity, H atoms have been omitted, and only the *ipso*-C atoms of the phenyl rings
are shown.

The ability of the bridging dppm ligand to support the dimetal core during chemical transformations is further illustrated in the following reactions. Treatment of $[Re_2Cl_4(dppm)_2]$ with $Na[BH_3CN]$ in methanol produced the first example of an aqua complex of rhenium(II); the aqua ligands in $[Re_2(dppm)_2(H_2O)_2(NCBH_3)_4]$ are in axial sites. When stirred in chloroform under air, $[Re_2Cl_4(dppm)_2]$ is oxidized to the paramagnetic complex

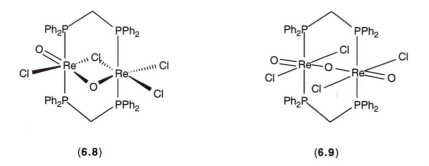

(6.8)　　　　　　　　　　　　　　(6.9)

$[Re_2Cl_3(O)(\mu\text{-}Cl)(\mu\text{-}O)(dppm)_2]$, **6.8**. Further oxidation with O_2 yields the rhenium(V) compound **6.9**. Both $[Re_2Cl_4(dppm)_2]$ and $[Re_2Br_4(dppm)_2]$ are extremely useful precursors for preparing derivatives which retain the $Re\equiv Re$ bond, since the $\{Re_2(dppm)_2\}^{4+}$ core is resistant to cleavage. A number of other ligands are capable of bridging the dimetallic core in a similar manner to dppm, for example 2-diphenylphosphinopyridine.

In the reactions described above, the dppe and dppm ligands favour bridging modes of bonding. This is not true for all didentate ligands however, and an early example was seen in $[Re_2Cl_5L_2]$ (L = $CH_3SCH_2CH_2SCH_3$) isolated from the reaction of $[Re_2Cl_8]^{2-}$ with L. The solid state structure reveals two chelating ligands bound to the same Re centre (Fig. 6.9) and an Re–Re bond length of 229 pm. The data are consistent with the presence of a $Re\equiv Re$ bond and the formulation of a mixed Re(II)Re(III) compound.

For further examples involving the $Re_2(dppm)_2$ unit, see: W. Wu, P.E. Fanwick and R.A. Walton (1997) *Inorg. Chem.*, **36**, 3810.

2-diphenylphosphinopyridine

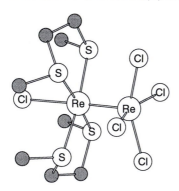

Fig. 6.9. The solid state structure of $[Re_2Cl_5L_2]$ (L = $CH_3SCH_2CH_2SCH_3$); H atoms have been omitted for clarity. The $\{ReCl_4\}$ and $\{ReS_4\}$ units are staggered.

6.5 [M$_2$(NMe$_2$)$_6$] (M = Mo or W)

The amido derivatives [Mo$_2$(NMe$_2$)$_6$] (Fig. 6.10) and [W$_2$(NMe$_2$)$_6$] contain metal-metal triple bonds (i.e. $\sigma^2\pi^4$). [Mo$_2$(NMe$_2$)$_6$] is made by treating MoCl$_3$ or MoCl$_5$ with LiNMe$_2$, while reaction of LiNMe$_2$ with WCl$_4$ yields [W$_2$(NMe$_2$)$_6$]. Both compounds adopt staggered configurations in the solid state (i.e. ethane-like, Fig. 6.10), and the Mo–Mo and W–W bond lengths are 221 and 229 pm respectively. The metal-nitrogen bonds are strengthened by π-contributions, and the orientations of the NMe$_2$ group are consistent with there being (metal *d*-nitrogen *p*)π-overlap. From Fig. 6.10, it is apparent that the Me groups either face into the M≡M bond (the *proximal* methyls) or away from it (the *distal* methyls) and in solution, NMR spectroscopic data indicate that there is exchange between the proximal and distal groups.

Both [Mo$_2$(NMe$_2$)$_6$] and [W$_2$(NMe$_2$)$_6$] are sensitive to air and moisture, and react with many organic solvents, but nonetheless are useful starting materials in this area of chemistry. Replacement of two amido groups by chloro ligands using Me$_3$SiCl gives [M$_2$Cl$_2$(NMe$_2$)$_4$] (M = Mo or W). These compounds again have ethane-like configurations (Fig. 6.11) and metal-metal triple bonds and, although very air- and moisture-sensitive, are important precursors for a variety of dimolybdenum or ditungsten species. Figure 6.12 gives representative examples of the uses of [M$_2$(NMe$_2$)$_6$] and [M$_2$Cl$_2$(NMe$_2$)$_4$] (M = Mo or W).

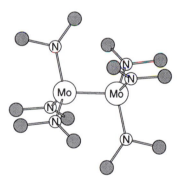

Fig. 6.10. The solid state structure of [Mo$_2$(NMe$_2$)$_6$] determined by X-ray diffraction; H atoms omitted for clarity.

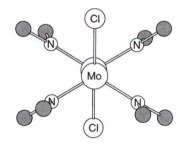

Fig. 6.11. The solid state structure (from X-ray diffraction) of [Mo$_2$Cl$_2$(NMe$_2$)$_4$], viewed down the Mo–Mo axis; H atoms omitted.

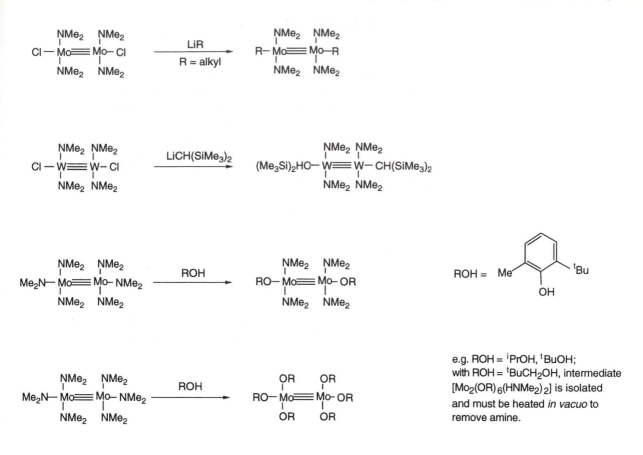

Fig. 6.12. Selected examples of reactions of [M$_2$(NMe$_2$)$_6$] and [M$_2$Cl$_2$(NMe$_2$)$_4$] (M = Mo or W).

One reaction of particular interest is the conversion of terminal dimethylamido ligands into bridging O$_2$CNMe$_2$ groups as a result of CO$_2$ insertion; Fig. 6.13 gives an example.

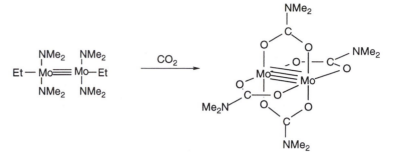

Fig. 6.13. Insertion of CO_2 into the Mo–N bonds in a Mo≡Mo dimethylamido complex results in an increase in the Mo–Mo bond order.

6.6 [$M_2(OR)_6$] (M = Mo or W; R = alkyl)

For reviews in this area, see:
M.H. Chisholm, D.M. Hoffman and J.C. Huffman (1985) *Chemical Society Reviews*, **14**, 69;
M.H. Chisholm (1995) *Chemical Society Reviews*, **24**, 79.

The chemistry of the M≡M bonded dimolybdenum and ditungsten alkoxides and their derivatives has been investigated in some detail over the last decade or so. The compounds [$Mo_2(OR)_6$] (R = e.g. ^iPr, ^tBu, CH_2^tBu) may be prepared from [$Mo_2(NMe_2)_6$] (see Fig. 6.12), and structural data for [$Mo_2(OCH_2^tBu)_6$] confirm the same staggered configuration as in [$Mo_2(NMe_2)_6$]. Although the compounds [$Mo_2(O^iPr)_6$], [$Mo_2(O^tBu)_6$] and [$Mo_2(OCH_2^tBu)_6$] are very air- and moisture-sensitive, volatile solids, they are valuable precursors to numerous derivatives (see below). The presence of bulky substituents is essential to the isolation of the discrete [$Mo_2(OR)_6$] species; even with sterically hindered groups, slow polymerization occurs, e.g. in hydrocarbons, [$Mo_2(OCH_2^tBu)_6$] slowly converts to [$Mo(OCH_2^tBu)_3]_n$. The tungsten compounds [$W_2(O^iPr)_6$] and [$W_2(O^tBu)_6$] are more difficult to handle than their molybdenum analogues. Although [$W_2(O^tBu)_6$] can be prepared by addition of ^tBuOH to [$W_2(NMe_2)_6$], an analogous reaction using ^iPrOH leads to [$W_4(O^iPr)_{14}H_2$], **6.10**, although in the presence of pyridine, the product is [$W_2(O^iPr)_6(py)_2$]. The reaction of EtOH with [$W_2(NMe_2)_6$] has been shown to give [$W_4(OEt)_{16}$], **6.11**. The systems are far from simple: e.g. in acetone at room temperature, [$W_2(O^tBu)_6(py)_2$] yields the tetranuclear species **6.12**, and the [$W_2(O^iPr)_6$] dimer (which can be made by adding excess ^iPrOH to [$W_2(O^tBu)_6$]) is in equilibrium with the corresponding tetramer, **6.13**.

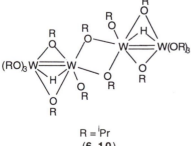

R = ^iPr
(**6.10**)

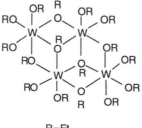

R=Et
(**6.11**)

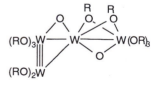

R = ^iPr
(**6.12**)

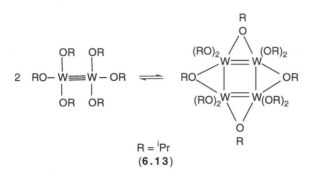

$$R = {}^{i}Pr$$
(6.13)

Compounds of the type $[M_2(OR)_6]$ readily undergo oxidative addition reactions, selected examples of which are given here. The reaction of $[Mo_2(O^iPr)_6]$ with Cl_2, Br_2 or I_2 gives $[Mo_2(O^iPr)_4(\mu\text{-}O^iPr)_2X_4]$ in which the halogen atoms are terminally bonded, two per metal; in going from $[Mo_2(O^iPr)_6]$ to $[Mo_2(O^iPr)_4(\mu\text{-}O^iPr)_2X_4]$, the metal-metal bond order is reduced from 3 to 1 (Mo–Mo = 273 pm). Addition of alkynes $RC\equiv CR$ to $[W_2(OR')_6]$ can lead to cleavage of the metal-metal bond and formation of the alkylidyne $[W(\equiv CR)(OR')_3]$ (e.g. Eqn. 6.5) although variations in products are observed depending upon the system; alkyne coupling may occur if the steric demands of the alkoxy group are less than those of O^tBu (e.g. Eqn. 6.6).

For a detailed account, see:
M.H. Chisholm (1996) *J. Chem. Soc., Dalton Trans.*, 1781.

$$[W_2(O^tBu)_6] + RC\equiv CR \xrightarrow{\text{273 K, hexane}} 2[W(\equiv CR)(O^tBu)_3] \qquad (6.5)$$

$$(R = Me, Et)$$

$$[W_2(O^iPr)_6] + 3MeC\equiv CMe \longrightarrow [W_2(O^iPr)_6(\mu\text{-}Me_4C_4)(\eta^2\text{-}MeC\equiv CMe)]$$

$$\textbf{(6.14)} \qquad (6.6)$$

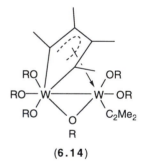

(6.14)

The addition of nitriles, RCN, to $[W_2(OR')_6]$ occurs to give **6.15** in, for example the case of R = Me and R' = CF_3CMe_2. However, for R = Me and R' = ^tBu, the reaction is accompanied by cleavage of the tungsten-tungsten and carbon-nitrogen bonds (Eqn. 6.7). Other variants on the reaction result from further substituent changes.

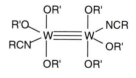

(6.15)

$$[W_2(O^tBu)_6] + MeC\equiv N \xrightarrow{\text{295 K, hexane}} [W(\equiv CMe)(O^tBu)_3] +$$

$$[W(\equiv N)(O^tBu)_3] \qquad (6.7)$$

The addition of CO to $[M_2(O^tBu)_6]$ (M = Mo or W) gives $[M_2(O^tBu)_4(\mu\text{-}O^tBu)_2(\mu\text{-}CO)]$ with a notable weakening of the carbonyl bond; $\nu_{CO} = 1575$ cm^{-1} in $[W_2(O^tBu)_4(\mu\text{-}O^tBu)_2(\mu\text{-}CO)]$. Again, the products depend on R and replacing O^tBu by O^iPr groups leads to the dimer $[\{W_2(O^iPr)_6(\mu\text{-}CO)\}_2]$ in

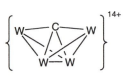

(6.16)

which each oxygen atom of the CO ligands bonds to a metal centre in the second W_2-unit. The reductive cleavage of CO has been observed in reactions with $[W_4(O^iPr)_{12}]$ (see **6.13**) in which the square W_4 framework transforms to a 'butterfly' unit containing a semi-interstitial carbon atom, **6.16**, the product being $[W_4(O^iPr)_{12}(C)(\mu\text{-}O)]$.

The area of metal alkoxide chemistry is a rich an exciting one, not least in the potential of these compounds to act as precursors to solid state materials.

7 Clusters with metal-metal bonds

7.1 Introduction

In an accompanying book in this series, *Metal-metal bonded carbonyl dimers and clusters*, we introduced cluster compounds containing low oxidation state, mid to late *d*-block metals with π-acceptor ligands. In this chapter, we discuss clusters containing higher oxidation state metals from the early to mid *d*-block with π-donor ligands. We shall focus on metal halide clusters and their derivatives, and in Chapter 8 we discuss polyoxometallates in which bridging oxo ligands support the metal cluster framework.

The definition of a 'cluster' is not uniform in textbooks but in this book, we consider a cluster as containing three or more metal centres in a closed framework. The three cluster cores which predominate in this chapter are the triangle (**7.1**), square-based pyramid (**7.2**) and octahedron (**7.3**). An *interstitial atom* is one which resides *fully* in the metal cage, for example, the notation μ_6-C is used to refer to a carbon atom sited within an octahedral M_6-cage. In structure **6.16**, we saw an example of a *semi-interstitial atom*, one which is partially surrounded by metal centres and is rather exposed.

C.E. Housecroft (1996) *Metal-metal bonded carbonyl dimers and clusters*, Oxford University Press, Oxford.

(7.1)   **(7.2)**

(7.3)

7.2 Trirhenium halides

The thermal decomposition of $ReCl_5$ leads to the formation of rhenium(III) chloride which exists as the trimer Re_3Cl_9 with structure **7.4**; the terminal chloro-ligands lie above and below the plane containing the Re and bridging Cl atoms. The Re–Re bond lengths are 248 pm, and each bond is assigned a bond order of two; it is useful to consider the molecule as possessing a formal $\{Re_3\}^{9+}$ core, one which appears in many of the derivatives discussed below. Interestingly, there is no technetium analogue of **7.4**. In the solid state, two-thirds of the terminal chlorine atoms in Re_3Cl_9 are involved in weak bridging interactions to rhenium atoms of adjacent molecules giving $Re(\mu$-Cl)$_2$Re' units (Re and Re' belong to different molecules); each Re atom is thus associated with five Cl atoms. The ability of the rhenium atoms in Re_3Cl_9 to interaction with electron donors plays an important role in its chemistry, and the addition of ligands to the in-plane sites shown in **7.5** gives rise to a series of derivatives. Examples include $[Re_3Cl_9(H_2O)_3]$ which can be isolated from Re_3Cl_9 in aqueous solution at 273 K, $[Re_3Cl_9(py)_3]$ which can be made from Re_3Cl_9 and pyridine, and $[Re_3Cl_9(CNMe)_3]$ which forms when Re_3Cl_9 is treated with MeNC. The addition of chloride ion leads to $[Re_3Cl_{10}]^-$, $[Re_3Cl_{11}]^{2-}$ or $[Re_3Cl_{12}]^{3-}$ depending upon the conditions of the reaction; for example, treatment of Re_3Cl_9 with an excess of CsCl in concentrated HCl gives $Cs_3[Re_3Cl_{12}]$ whereas reaction of Re_3Cl_9 with aqueous HCl in the presence of Ph_4AsCl leads to $[Ph_4As]_2[Re_3Cl_{11}]$.

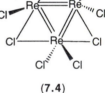

(7.4)

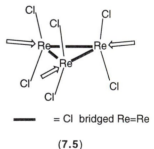

—— = Cl bridged Re=Re

(7.5)

Rhenium(III) bromide (prepared by dehydration of $ReBr_3 \cdot {}^2/_3 H_2O$) is also trinuclear and forms a series of complexes e.g. $[Re_3Br_{12}]^{3-}$, which resemble their chloro analogues. Rhenium(III) iodide is thermally less stable than the chloride or bromide, and decomposes slowly *in vacuo*. It can be synthesized by reaction of $ReCl_3$ with an excess of BiI_3 at 580 K.

7.3 Halides of the group 5 and 6 metals

In Section 4.3, we noted that binary halides of niobium and tantalum (e.g. NbF_5 and $TaCl_5$) are not monomeric but possess halide-bridged structures. Reduction of these halides with cadmium at red heat produces systems which feature $[M_6X_{12}]^{2+}$ clusters, **7.6**. These units are present in halides of formulae M_6X_{14} or M_6X_{15}, the difference in formula arising from the

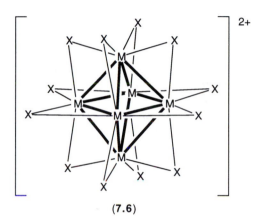

(7.6)

Counting electrons in the $[M_6X_{12}]^{2+}$ (M = Nb, Ta) cluster core: see page 70.

manner in which the clusters are connected in the solid state lattice. The formula M_6X_{14} can be usefully rewritten as $[M_6X_{12}]X_{4/2}$, signifying the association of each $[M_6X_{12}]^{2+}$ cluster with four bridging X^- ligands generating an overall sheet structure. Similarly, M_6X_{15} can be rewritten as $[M_6X_{12}]X_{6/2}$ which indicates sharing of six halide ions per $[M_6X_{12}]^{2+}$ cluster to give a three-dimensional lattice. There are several important points to note about the $[M_6X_{12}]^{2+}$ clusters:

- the formal oxidation state of the metal is non-integral;
- M_6X_{14} compounds are diamagnetic, indicating the presence of metal-metal bonding;
- M_6X_{15} compounds are paramagnetic, with magnetic moments that correspond to one unpaired electron per metal centre, and thus, some metal-metal bonding must be present.

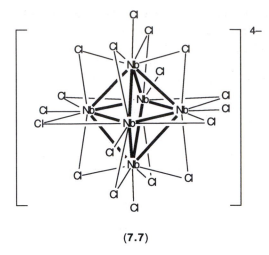

(7.7)

Discrete clusters also exist, and include $[Nb_6Cl_{18}]^{4-}$, **7.7**, formed (as the potassium salt) by the reaction of Nb_6Cl_{14} with KCl at 920 K. Oxidation of $[Nb_6Cl_{18}]^{4-}$ with I_2 (or air oxidation of Nb_6Cl_{14} in ethanol saturated with gaseous HCl) produces $[Nb_6Cl_{18}]^{3-}$, whereas oxidation of $[Nb_6Cl_{18}]^{4-}$ using Cl_2 leads to the formation of $[Nb_6Cl_{18}]^{2-}$. The corresponding bromide analogues are also known, as are examples in which the terminal halo ligands have been replaced by other donors, e.g. conversion of $[Nb_6Br_{18}]^{4-}$ to $[Nb_6Br_{12}(N_3)_6]^{4-}$.

Similar reactions can be carried out starting from Ta_6Cl_{14}. For example, treatment with KCN in methanol, followed by HCN, produces $H_4[Ta_6Cl_{12}(CN)_6]$ in which the *C*-bonded cyano ligands occupy the six terminal sites (structure **7.8a**). The oxidation by air of a solution of $[Ta_6Cl_{12}]Cl_2\cdot8H_2O$ in methanol followed by the addition of $[Et_4N]OH$ leads to the formation of $[Et_4N]_2[Ta_6Cl_{12}(OH)_6]$ containing anion **7.8b**. A

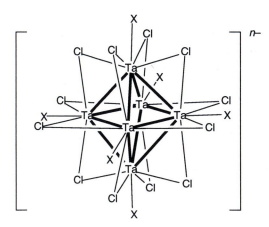

(7.8a) X = CN, $n = 4$

(7.8b) X = OH, $n = 2$

(7.8c) X = CF$_3$SO$_3$, $n = 2$

useful starting material for other derivatives is $[Ta_6Cl_{12}(OSO_2CF_3)_6]^{2-}$ (**7.8c**) which contains labile trifluoromethanesulfonate ligands; it can be made by treating $[Ta_6Cl_{18}]^{2-}$ with CF_3SO_3H. Reaction of $Ta_6Cl_{14}\cdot 8H_2O$ with an excess of PEt_3 leads to the formation of a mixture of *trans*- and *cis*-$[Ta_6Cl_{14}(PEt_3)_4]$ (there are two terminal Cl ligands which are mutually *trans* or *cis* with respect to the octahedral metal core). Oxidation of these isomers using $[NO][BF_4]$ produces the *trans*- and *cis*-isomers of $[Ta_6Cl_{14}(PEt_3)_4]^{n+}$ ($n = 1$ or 2), isolated as the $[BF_4]^-$ salts.

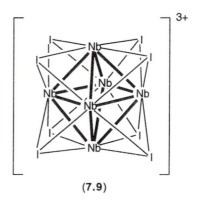

(**7.9**)

The structural features common to niobium and tantalum halide clusters are generally an octahedral metal core supported by edge-bridging and terminal ligands, the latter sometimes being involved in bridging to another cluster unit. Exceptions to these generalities include Nb_6I_{11} and $[Nb_6SBr_{17}]^{3-}$. The former is paramagnetic and contains μ_3-iodo ligands facially capping an octahedral core, and is informatively written as $[Nb_6I_8]I_{6/2}$. The $[Nb_6I_8]^{3+}$ unit is shown in structure **7.9** and in the solid state of Nb_6I_{11}, these cluster units are connected by bridging iodine atoms, six being associated with each cluster, to give a three-dimensional lattice. The anion $[Nb_6SBr_{17}]^{3-}$ is particularly unusual in containing a sulfur-centred, trigonal prismatic Nb_6-cluster **7.10**; the rubidium salt $Rb_3[Nb_6SBr_{17}]$ is formed when niobium is heated with RbBr, $NbBr_5$ and sulfur in a sealed tube at ≈ 1100 K.

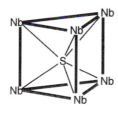

(**7.10**)

The molybdenum and tungsten dichlorides, dibromides and diiodides are hexameric (i.e. M_6X_{12}) and, in the solid state, contain halide-bridged $[M_6X_8]^{4+}$ clusters which are structurally analogous to **7.9** with face-capping (μ_3) halide ligands. The overall formulae of the dihalides can be written as $[M_6X_8]X_2X_{4/2}$ signifying connection of the cluster units into sheets; two halides are terminally attached and remain unshared in the solid state lattice. The halides are diamagnetic and this can be rationalized in terms of the presence of metal-metal bonds in the octahedral cluster framework. The electrons available for bonding within the $[M_6X_8]^{4+}$ core can be allocated as follows, using a model which ignores oxidation states:

- each Mo or W atom has 6 valence electrons;
- each X atom provides 1 electron for bonding;
- allowing for the overall charge of 4+, the total number of valence electrons available is 40, or 20 pairs;

For a detailed discussion of the bonding in metal clusters with π-donor ligands, see:
Z. Lin and M.-F. Fan (1997)
Structure and Bonding, **87**, 35.

- to form the M–X 2c-2e localized bonds, 8 pairs are needed, leaving 12 pairs for M–M bonding;
- the octahedron has 12 edges, and so each edge can be considered to be a single bond.

2c-2e = 2-centre 2-electron

Now let us return to the $[M_6X_{12}]^{2+}$ (M = Nb or Ta) clusters. Each of $[M_6X_{12}]^{2+}$ (M = Nb or Ta) and $[M_6X_8]^{4+}$ (M = Mo or W) possesses the same number (40) of valence electrons. In the $[M_6X_{12}]^{2+}$ clusters, 12 electron pairs are needed to form localized 2c-2e M–X bonds, leaving 8 pairs for M–M bonding. Allocated over the 12 edges of the octahedron, this gives a bond order of $^2/_3$ per M–M edge.

Note: irrespective of the type of M–Cl interaction (terminal, μ or μ_3), only one electron from the cluster core need be allocated. (Why is this?)

Before looking at representative reactions of the Mo_6Cl_{12}, we should note that a second form of the dichloride is produced when $[Mo_2(O_2CMe)_4]$ reacts with dry HCl at 570 K; this brown solid is much more reactive than the yellow hexamer.

In most of its reactions, Mo_6Cl_{12} retains the $[Mo_6Cl_8]^{4+}$ core. It reacts with $[Et_4N]Cl$ in dilute HCl to form $[Et_4N]_2[Mo_6Cl_{14}]$; a large number of salts containing the diamagnetic $[Mo_6Cl_{14}]^{2-}$ ion (**7.11**) are known.

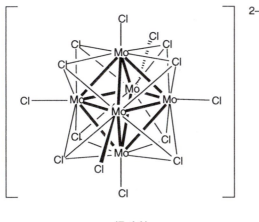

(**7.11**)

Replacement of the six terminal Cl^- ligands by triflate groups (Eqn. 7.1) introduces labile ligands, making $[Mo_6Cl_8(OSO_2CF_3)_6]^{2-}$ a useful precursor for other derivatives.

Triflate = trifluorosulfonate

$[CF_3SO_3]^-$

$$[Bu_4N]_2[Mo_6Cl_{14}] + 6Ag(CF_3SO_3) \xrightarrow{CH_2Cl_2} [Bu_4N]_2[Mo_6Cl_8(OSO_2CF_3)_6]$$

$$+ 6AgCl \qquad (7.1)$$

Alkoxide ligands can also be introduced into the terminal sites, e.g. to give $[Mo_6Cl_8(OR)_6]^{2-}$ (R = Me or Ph), and the reaction between Mo_6Cl_{12} and $[Bu_4N][NCS]$, NaOH and HBF_4 in methanol produces $[Mo_6Cl_8(NCS-N)_6]^{2-}$. On the other hand, reaction of Mo_6Cl_{12} with the phosphine PBu_3 in

thf solution results in the addition of just two terminal ligands and the formation of [$Mo_6Cl_{12}(PBu_3)_2$].

An interesting reaction of Mo_6Cl_{12} occurs in an $AlCl_3/KCl/BiCl_3$ melt; the hexanuclear core is degraded to a pentanuclear framework with square-based pyramidal structure. The [Mo_5Cl_{13}]$^{2-}$ ion (**7.12**) has been isolated as the tetrabutylammonium salt.

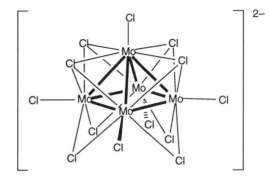

(7.12)

(7.13)

7.4 Cluster halides of zirconium

When $ZrCl_4$ is heated with zirconium powder and elemental carbon in a sealed tube at temperatures ≥ 1000 K, the product is [$Zr_6Cl_{14}C$]. Relatively early studies concerning reactions of $ZrCl_4$ and Zr under similar conditions reported the formation of cluster zirconium chlorides, structurally similar to the niobium and tantalum species discussed in the last section, but more recent data have shown the presence of an interstitial atom (e.g. carbon). The crucial role of an interstitial atom is that it provides valence electrons for the stabilization of an otherwise electron-poor compound from *within the cage* without imposing steric demands among the ligands around the outside of the cage. Members of a series of related species have now been prepared and structurally characterized. The reaction of zirconium, carbon, $ZrBr_4$ and CsBr at 1100 K in a sealed tantalum tube for 3 weeks yields Cs_3[$Zr_6Br_{15}C$]; the anion [$Zr_6Br_{15}C$]$^{3-}$ can be rewritten in the form of [$\{Zr_6Br_{12}C\}Br_{6/2}$]$^{3-}$ to provide more structural information (twelve μ-Br and six *shared* terminal Br, see the discussion in Section 7.2). The boron-centred cluster [$Zr_6Cl_{18}B$]$^{5-}$ is obtained as the rubidium salt when boron and zirconium powders are heated with $ZrCl_4$ and RbCl at 1120 K in a sealed tube, and the compound K_2[$Zr_6Cl_{15}B$] has been obtained from the reaction between $ZrCl_4$, KCl, B and Zr at 1220 K. By using the room temperature ionic liquid $AlCl_3$ with 1-ethyl-3-methylimidazolium chloride, [**7.13**]Cl, it has been possible to extract and isolate the salt [**7.13**]$_5$[$Zr_6Cl_{18}B$] (Fig. 7.1) from the solid state Rb_5[$Zr_6Cl_{18}B$]; use of ionic liquid media has proved to be a successful means of isolating clusters of this type without their undergoing oxidation.

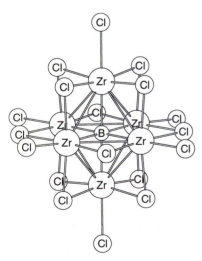

Fig. 7.1. The structure of [$Zr_6Cl_{15}B$]$^{5-}$ determined by X-ray diffraction for the salt [**7.13**]$_5$[$Zr_6Cl_{15}B$].

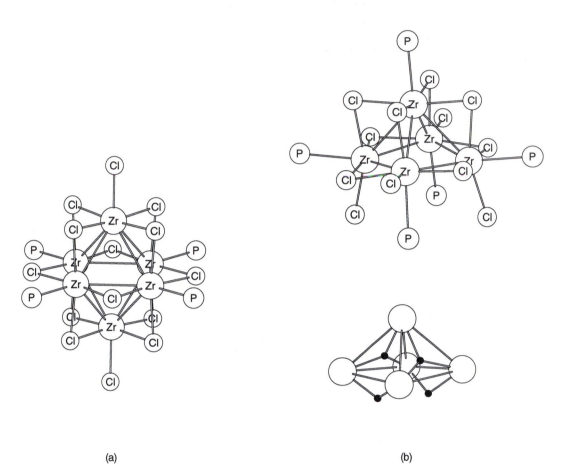

Fig. 7.2. The structures of (a) $[Zr_6Cl_{14}(PMe_3)_4]$ and (b) $[Zr_5Cl_{12}(\mu\text{-}H)_2(\mu_3\text{-}H)_2(PMe_3)_5]$ determined by X-ray diffraction. In (b), the top view shows the molecule without cluster H atoms; the lower view shows the Zr_5-core and the cluster H atoms. All P substituents are omitted for clarity.

While a certain stabilizing influence, the presence of the interstitial atom may not be essential as has been shown by the isolation of discrete clusters such as $[Zr_6Cl_{14}(PMe_3)_4]$ (Fig. 7.2a). The synthesis of this and related derivatives is of importance not only for the lack of the interstitial atom but also for the fact that the cluster is assembled at far lower temperatures than the preparations so far described; reduction of $ZrCl_4$ by Bu_3SnH followed by treatment with PMe_3 leads to $[Zr_6Cl_{14}(PMe_3)_4]$. A rare example of a pentazirconium cluster, $[Zr_5Cl_{12}(\mu\text{-}H)_2(\mu_3\text{-}H)_2(PMe_3)_5]$ (Fig. 7.2b) has also been isolated from the reduction of $ZrCl_4$ by Bu_3SnH and reaction with PMe_3. This combination of reagents (also with different phosphine ligands)

Work in this area has provided intriguing structural and bonding problems, and is a current topic of interest; see for example:
F.A. Cotton and W.A. Wojtczak, *Inorg. Chim. Acta.* (1994) **223**, 93;
L. Chen, F.A. Cotton and W.A. Wojtczak, *Inorg. Chem.* (1996) **35**, 2988; *Inorg. Chem.* (1997), **36**, 4047.

has provided a general method of preparing a wide range of hexazirconium clusters with hydrido ligands, e.g. $[Zr_6Cl_{14}(PR_3)H_4]$ (R = Me, Et or ^nPr), $[Zr_6Cl_{18}H_4]^{3-}$ and $[Zr_6Cl_{18}H_5]^{4-}$. The system is extremely complicated and varying choices of solvents result in isolation of different cluster species. More recently, it has been found that following the reduction step with treatment of $[Ph_4P]Cl$ (instead of PR_3) gives an improved route to $[Zr_6Cl_{18}H_4]^{3-}$ (as the $[Ph_4P]^+$ salt). These studies are recent, and have brought about a reappraisal of the nature of $[Zr_6Cl_{12}(PMe_2Ph)_6]$. This has been shown by X-ray diffraction to possess the structure shown in Fig. 7.3 comprising a Zr_6 cage devoid of an interstitial atom. Initial studies provided only very small samples of the compound, but improved yields are now available using the reduction of $ZrCl_4$ by $^nBu_3SnH$, followed by addition of PMe_2Ph and further reduction using sodium amalgam. It appears possible that '$[Zr_6Cl_{12}(PMe_2Ph)_6]$' may in fact contain an interstitial hydride.

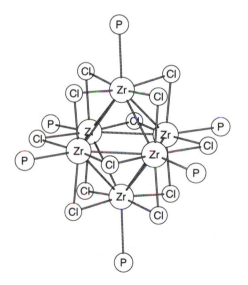

Fig. 7.3.   The structure of $[Zr_6Cl_{12}(PMe_2Ph)_6]$ determined by X-ray diffraction; the phenyl and methyl groups have been omitted for clarity.

8 Polyoxometallates of molybdenum and tungsten

8.1 Introduction: condensation of $[MO_4]^{2-}$ units

In Section 3.6, we discussed the formation of oxoanions of molybdenum and tungsten in solution and described how condensation of $[MO_4]^{2-}$ (M = Mo or W) may occur. Equation 8.1 shows the condensation of $[MoO_4]^{2-}$ to give a polyoxoanion or *polyoxometallate*, sometimes abbreviated to POM.

$$7[MoO_4]^{2-} + 8H^+ \rightarrow [Mo_7O_{24}]^{6-} + 4H_2O \qquad (8.1)$$

Equation 8.1 illustrates that the condensation of $[MoO_4]^{2-}$ ions requires H^+; condensation processes of molybdates and tungstates are highly pH dependent, and reaction 8.1 takes place at pH≈5. The process involves the sharing of oxygen atoms between metal centres, and concomitant changes in the coordination environments of the metal centres. The mononuclear $[MoO_4]^{2-}$ ion, **8.1**, is tetrahedral, but in $[Mo_7O_{24}]^{6-}$, each molybdenum centre is octahedrally sited. Before looking at the structure of $[Mo_7O_{24}]^{6-}$, we need to introduce a method of representing the structures of polyoxometallates. This is similar to that used to represent silicates, the focus of attention in structural diagrams being the arrangement of the oxygen atoms.

Figure 8.1a shows how an octahedral MO_6 unit is usually represented: only the octahedron defined by the oxygen atoms is drawn as shown on the right of the diagram. If *two* MO_6 units share an oxygen atom, this corresponds to the formation of an M–O–M bridge; Fig. 8.1b illustrates how the formation of two such bridges between two MO_6 units may be represented. Similarly, a μ_3-O atom is present when three metal centres share a common oxygen atom, and so on. Most commonly, the octahedral units are represented simply, with the back faces omitted, e.g. **8.2**, but in this book, we shall retain all oxygen vertices for clarity.

The structure of $[Mo_7O_{24}]^{6-}$ is shown schematically in Fig. 8.2a and to help with the interpretation of the diagram, Fig. 8.2b shows the formal construction of the structure in two steps. For comparison, Fig. 8.3 shows a 'ball-and-stick' diagram of the structure, drawn in the same projection as that in Fig 8.2a. Notice that, although the schematic representations indicate

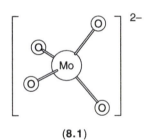

(8.1)

All vertices   Back vertex
shown.   omitted.

(8.2)

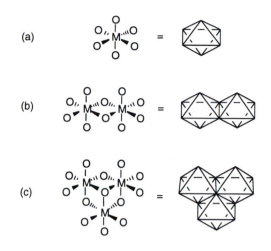

(a)

(b)

(c)

Fig. 8.1. Representations of structural units in polyoxometallates: (a) an octahedral MO_6 unit; (b) two octahedral MoO_6 units sharing two oxygen atoms; (c) three MO_6 units sharing four oxygen atoms, only one of which is common to all three units.

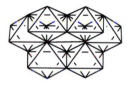

(a)

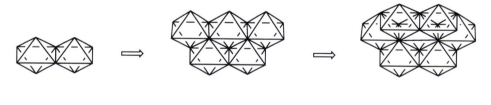

(b)

Fig. 8.2. (a) The structure of $[Mo_7O_{24}]^{6-}$. (b) Building up the structure of $[Mo_7O_{24}]^{6-}$ by interconnection of seven Mo_6 octahedra; the structure contains twelve terminal oxygen atoms, eight μ-O atoms, two μ_3-O atoms, and two μ_4-O atoms . Where are the different types of oxygen atoms positioned in the structure of $[Mo_7O_{24}]^{6-}$? (To check the answer, look at Fig. 8.3.)

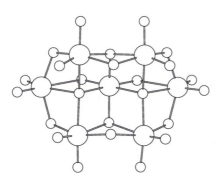

For detailed discussion of the structural families of polyoxometallates, see: A.F. Wells (1984) *Structural Inorganic Chemistry*, 5th ed., Clarendon Press, Oxford, Chapter 11.

Fig. 8.3. The structure of $[Mo_7O_{24}]^{6-}$ determined by X-ray diffraction for the $[H_3N(CH_2)_2NH_2(CH_2)_2NH_3]^{3+}$ salt.

regular octahedral units, the actual environments of the metal centres in the solid state are distorted.

8.2 Polyoxomolybdates and -tungstates

As we have seen, at pH≈5, aqueous solutions of molybdates contain the $[Mo_7O_{24}]^{6-}$ ion; no lower nuclearity species are present. However, by careful control of pH or by working in non-aqueous media, it is possible to isolate other polyoxoanions of molybdenum as crystalline salts.

A range of salts containing the $[Mo_6O_{19}]^{2-}$ ion is known, and Fig. 8.4 shows that the ion consists of six octahedral MoO_6 units with one oxygen atom common to all six units; the notation for the latter is μ_6-O. The symmetrical structure of $[Mo_6O_{19}]^{2-}$ is adopted by a number of other ions, e.g. $[W_6O_{19}]^{2-}$ and $[Nb_6O_{19}]^{8-}$, and is known as the *Lindqvist structure*.

Structural characterization of salts of $[Mo_8O_{26}]^{4-}$ reveal that three isomers (the α-, β- and γ-forms) exist. Of particular interest is the γ-isomer, which contains two five-coordinate molybdenum centres; this is compared with the structure of the β-isomer in Fig. 8.5.

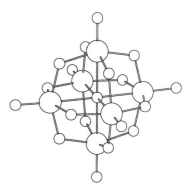

Fig. 8.4. The structure of $[Mo_6O_{19}]^{2-}$ determined by X-ray diffraction for the $[Bu_4N]^+$ salt.

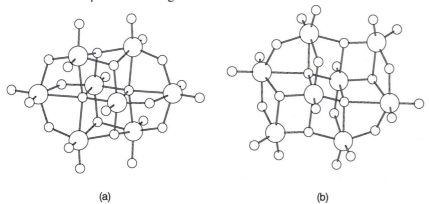

(a)   (b)

Fig. 8.5. The structures (determined by X-ray diffraction) of (a) β-$[Mo_8O_{26}]^{4-}$ in the compound $[Et_3NH]_3[H_3O][Mo_8O_{26}]$, and (b) γ-$[Mo_8O_{26}]^{4-}$ in the $[Me_3N(CH_2)_6NMe_3]^{2+}$ salt.

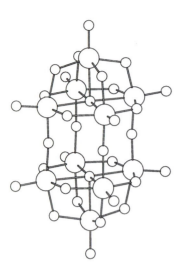

Fig. 8.6. The structure of the $[W_{10}O_{32}]^{4-}$ ion determined by X-ray diffraction for the $[Bu_3NH]^+$ salt; note the relationship to the Lindqvist structure (Fig. 8.4).

The pattern of formation of polyoxotungstates in aqueous solution does not mimic that of molybdenum, although the lowest nuclearity species is $[W_7O_{24}]^{6-}$, analogous to the molybdate shown in Fig. 8.3. In solution, there is a complex system of equilibria involving $[W_7O_{24}]^{6-}$ and W_{10}, W_{11} and W_{12} polyoxoanions. Again, careful control of pH or other conditions (e.g. use of non-aqueous media) allow the crystallization of salts of different polyoxometallates, some of which do not exist in aqueous solution. An example is $[W_{10}O_{32}]^{4-}$, the structure of which is shown in Fig. 8.6.

8.3 Heteropolyoxometallates: Keggin and Dawson structures

Heteropolymetallates of molybdenum and tungsten are those which contain atoms other than Mo and O, or W and O. The heteroatoms are commonly from the *p*-block, e.g. B, Si, Ge, P(V), As(V) and S. The first family of heteropolyoxometallates that we consider are those of general formula $[XM_{12}O_{40}]^{n-}$ (M = Mo, W; e.g. X = P, n = 3; X = Si, n = 4). The original member of this family, $[PMo_{12}O_{40}]^{3-}$, was first reported in 1826 by Berzelius, but it was not until 1933 that its structure was determined by J.F. Keggin using X-ray diffraction. The family is now known as the *α-Keggin anions* (α being used to distinguish the structure-type from other structural isomers). Equations 8.2 and 8.3 exemplify typical syntheses, and the structure of a representative α-Keggin anion is shown in Fig. 8.7. The heteroatom is tetrahedrally coordinated by four oxygen atoms at the centre of

$$[SiO_3]^{2-} + 12[WO_4]^{2-} + 22H^+ \rightarrow [SiW_{12}O_{40}]^{4-} + 11H_2O \qquad (8.2)$$

$$[HPO_4]^{2-} + 12[WO_4]^{2-} + 23H^+ \rightarrow [PW_{12}O_{40}]^{3-} + 12H_2O \qquad (8.3)$$

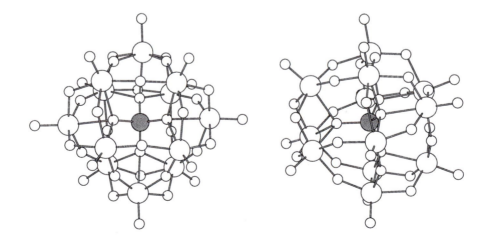

Fig. 8.7. An example of an α-Keggin anion: two views of $[SiMo_{12}O_{40}]^{4-}$; the structure was determined by X-ray diffraction for the guanidinium salt.

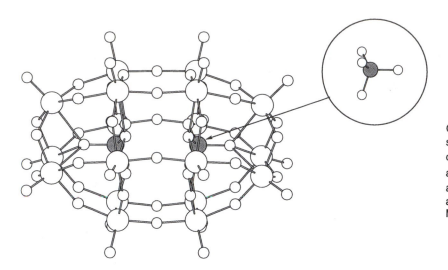

Compare Fig. 8.8 with the right-hand side of Fig. 8.7 to understand how condensation of two α-Keggin anions leads to the α-Dawson anion; condensation is accompanied by formal loss of six MO_3 units.

Fig. 8.8. The α-Dawson structure-type: the inset emphasizes the tetrahedral coordination environment of the heteroatom. The figure shows the structure of $[H_3S_2Mo_{18}O_{62}]^{5-}$ formed by reduction of the α-Dawson anion $[S_2Mo_{18}O_{62}]^{4-}$; reduction is accompanied by some bond length changes but no gross structural change (see text).

the oxometallate cage, and each metal atom is in an octahedral environment. By writing the general formula of an α-Keggin anion in the form $[\{XO_4\}M_{12}O_{36}]^{n-}$, the coordination mode of the heteroatom X can be emphasized.

The condensation of two α-Keggin anions by the extrusion of six $\{MO_3\}$ groups leads to the formation of an α-*Dawson anion* of general formula $[X_2M_{18}O_{62}]^{n-}$ or $[\{XO_4\}_2M_{18}O_{54}]^{n-}$. This family includes $[P_2Mo_{18}O_{62}]^{6-}$, $[P_2W_{18}O_{62}]^{6-}$, $[As_2Mo_{18}O_{62}]^{6-}$, $[As_2W_{18}O_{62}]^{6-}$ and $[S_2Mo_{18}O_{62}]^{4-}$. The anions $[P_2Mo_{18}O_{62}]^{6-}$ and $[As_2Mo_{18}O_{62}]^{6-}$ are formed readily in solutions of molybdophosphates or arsenates, while heating $[WO_4]^{2-}$ and As_2O_5 in aqueous solution yields $[As_2W_{18}O_{62}]^{6-}$, and the reaction between $[WO_4]^{2-}$ and aqueous phosphoric acid produces $[P_2W_{18}O_{62}]^{6-}$. As with the Keggin structure-type, several isomers of the Dawson cage are known. The general structure-type for an α-Dawson anion is illustrated in Fig. 8.8; this shows the structure of the reduced anion $[H_3S_2Mo_{18}O_{62}]^{5-}$, formed by treating the Dawson anion $[S_2Mo_{18}O_{62}]^{4-}$ with PPh_3 in acetonitrile. While the basic structure of the polyoxometallate cage is retained, reduction is accompanied by an increase in the $Mo\cdots Mo$ separations around the equatorial belts, and a decrease in the Mo–O–Mo bond lengths in the bridges that connect the two halves of the cage. Reduction of the α-Keggin anion $[PMo_{12}O_{40}]^{3-}$ to $[HPMo_{12}O_{40}]^{4-}$ has also been studied and again, the basic structure is retained as electrons are added to the cage.

Remember: protonation does not add (or remove) electrons from the polyoxometallate cage.

Electrochemical reduction is another means of adding electrons, (e.g. the conversion of $[As_2W_{18}O_{62}]^{6-}$ to $[As_2W_{18}O_{62}]^{7-}$) and up to six electrons may be reversibly added, reducing M(VI) to M(V) centres. Reduction of Keggin and Dawson anions leads to the so-called 'heteropoly blues', named because of their intense blue colour, due to intervalence charge transfer transitions

8.4 Lacunary heteropolyoxometallates

We have already emphasized that the formation of a particular polyoxometallate is crucially dependent on the pH of the solution. Under appropriate conditions, it is possible to isolate polyoxometallates with 'incomplete' cages, the so-called *lacunary* species. For example, whereas $[PW_{12}O_{40}]^{3-}$ is stable in aqueous solution at pH$\approx$1, a change to pH$\approx$2 leads to the formation of $[PW_{11}O_{39}]^{7-}$ (or protonated form thereof) i.e. formal loss of a $[WO]^{4+}$ fragment. Two series of lacunary anions are known: $[XM_{11}O_{39}]^{n-}$ and $[XM_9O_{34}]^{n-}$, both derived from the Keggin anion. The removal of metal units from the closed Keggin cage (Fig. 8.7) leaves the lacunary polyoxometallate with available oxygen donor atoms and, thus, it may function as a polydentate ligand. Numerous complexes have been prepared and characterized, with the lacunary ligands coordinating to fragments involving *p*- or *d*-block metals, e.g. $[(SiW_9O_{37})(SnPh)_3]^{7-}$ (Fig. 8.9a), $[(SiW_9O_{34})_2(PhSnOH)_3]^{14-}$ (Fig. 8.9b) $[(SiW_{11}O_{39})(SnPh)]^{5-}$, $[Co_4\{AsW_9O_{34}\}_2(H_2O)_2]^{10-}$, $[(PW_{11}O_{39})\{Ti(\eta^5\text{-}C_5H_5)\}]^{4-}$ and $[(PW_{11}O_{39})\{Zr(\eta^5\text{-}C_5Me_5)\}]^{4-}$ and $[(PW_{11}O_{39})Rh_2(O_2CMe)_2(dmso)_2]^{5-}$.

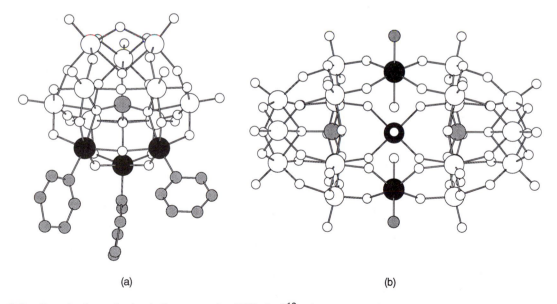

(a)   (b)

Fig. 8.9. Complex formation by the lacunary anion $[SiW_9O_{34}]^{10-}$: the structures (determined by X-ray diffraction) of (a) $[(SiW_9O_{37})(SnPh)_3]^{7-}$ (oxygen is gained during the reaction), and (b) $[(SiW_9O_{34})_2(PhSnOH)_3]^{7-}$ (only the *ipso*-carbon of each Ph ring is shown and H atoms are omitted). Tin atoms are shown in black.

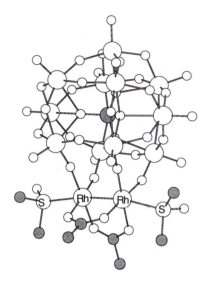

Fig. 8.10. The solid state structure of $[(PW_{11}O_{39})Rh_2(O_2CMe)_2(dmso)_2]^{5-}$
(determined for the guanidinium salt); compare the rhodium coordination environment
with that in structure **6.5**.

(Fig. 8.10). This last complex is of interest because it combines the chemistry of the rhodium(II) dimers that we discussed on page 56 with that of polyoxometallate clusters. The chelation of a lacunary anion to another *metal oxo* centre has the effect of 'exchanging' a metal centre in the Keggin ion, e.g. $[PMo_{11}VO_{40}]^{4-}$, although exact synthesis of a given ion may not follow this formal description.

8.5 Derivatives of polyoxometallates

We have already seen that complex formation by lacunary anions is one means of incorporating new groups into Keggin-derived anions. In this section, we give selected examples of ways of derivatizing polyoxo-metallates, and we focus on derivatives of $[M_6O_{19}]^{2-}$ (M = Mo or W).

The reaction of $[WO_4]^{2-}$ and $[(\eta^5\text{-}C_5R_5)_2TiCl_2]$ (R = H or Me) leads to the formation of $[(\eta^5\text{-}C_5R_5)TiW_5O_{18}]^{3-}$, and with $[Mo_2O_7]^{2-}$ as precursor, the products are $[(\eta^5\text{-}C_5R_5)TiMo_5O_{18}]^{3-}$ (R = H or Me). Figure 8.11 shows the structure of $[(\eta^5\text{-}C_5H_5)TiMo_5O_{18}]^{3-}$ and illustrates the retention of the Lindqvist-type structure of the parent $[Mo_6O_{19}]^{2-}$. The incorporation of the organometallic fragment has the effect of making the polyoxometallate more reactive in respect of 'surface reactions', i.e. the terminal oxygen atoms are able to function as donors. Thus, $[(\eta^5\text{-}C_5H_5)TiMo_5O_{18}]^{3-}$ reacts with $[Mn(CO)_3(NCMe)_3]^+$ to form $[(\eta^5\text{-}C_5H_5)TiMo_5O_{18}\{Mn(CO)_3\}]^{2-}$, in which the $Mn(CO)_3$ fragment is bound to the polyoxometallate cluster by three Mo-O-Mn interactions. Substitution of Nb(V) or V(V) in place of Mo(VI) or W(VI) has a similar effect, and examples of derivatives that exhibit 'surface-

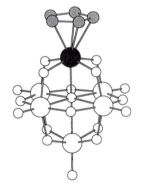

Fig. 8.11. The structure of $[(\eta^5\text{-}C_5H_5)TiMo_5O_{18}]^{3-}$ determined by X-ray diffraction for the $[Bu_4N]^+$ salt. The titanium centre is shown in black.

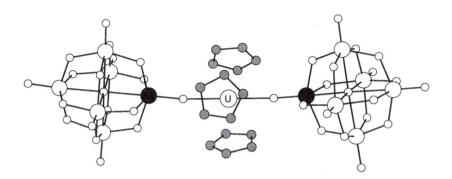

Fig. 8.12. The structure of $[(NbW_5O_{19})_2U(\eta^5-C_5H_5)_3]^{5-}$, determined by X-ray diffraction for the $[^nBu_4N]^+$ salt; for clarity, H atoms have been omitted and U–C bonds have not been drawn in. The niobium atoms are shown in black.

nbd = 2,5-norbornadiene

bound' groups include $[Nb_2W_4O_{19}\{Re(CO)_3\}]^{3-}$ (in which the rhenium centre is attached to the Lindqvist anion through three M–O–Re bridges), $[(Nb_2W_4O_{19})_2\{\mu\text{-}Rh(nbd)\}_5]^{3-}$ (in which five Rh centres are held by O–Rh–O bridges between two $[Nb_2W_4O_{19}]^{4-}$ cages) and $[(\eta^5\text{-}C_5Me_5)TiW_5O_{18}\{Ru(\eta^6\text{-}C_6H_6)\}]^-$ (in which the Ru centre is attached to the cage through three W–O–Ru bridges). A further example is the reaction of $[NbW_5O_{19}]^{3-}$ with $[(\eta^5\text{-}C_5H_5)_3UCl]$ in dichloromethane to yield the novel ion $[(NbW_5O_{19})_2U(\eta^5\text{-}C_5H_5)_3]^{5-}$ the structure of which is shown in Fig. 8.12. It is important to note that working in *non-aqueous media* (e.g. dichloromethane or acetonitrile) is usual in these systems and choice of counterion (e.g. $[^nBu_4N]^+$) for the polyoxometallates is important in order to obtain a salt which is soluble in the organic solvent.

The assembly of systems containing one or more $[M_5O_{18}]^{6-}$ (M = Mo or W) ligands, structurally derived from a Lindqvist anion is illustrated by the following examples. The reaction of $[WO_4]^{2-}$ with $[PdCl_4]^{2-}$ at 350 K in aqueous solution at pH 5 leads to the formation of $[Pd_2(W_5O_{18})_2]^{8-}$. Each Pd(II) centre is in the expected square planar environment, coordinated by four oxygen atoms, two from each polyoxometallate ligand. In the reaction of $[WO_4]^{2-}$ and $Eu(NO_3)_3$ in aqueous media at pH 7-7.5, the complex $[Eu(W_5O_{18})_2]^{9-}$ is formed. The two $[W_5O_{18}]^{6-}$ ligands act as tetradentate ligands with the europium(III) centre sandwiched between them in a square antiprismatic coordination environment.

Replacing terminal oxygen atoms in polyoxometallate ions by organic functionalities is one method by which the cage may be prepared for further reaction; for example, cages may be coupled. In the case of $[M_6O_{19}]^{2-}$, the six terminal M=O bonds bear an orthogonal relationship to one another as structure **8.3** (which views the Lindqvist structure down a terminal O=M bond) shows. Functionalization, therefore, could ultimately lead to the construction of an ordered array of connected cages. The incorporation of imido groups has received recent attention. The reaction between $[Mo_6O_{19}]^{2-}$ and $Ph_3P=NC_6H_4Me$, **8.4**, in anhydrous pyridine at 358 K yields

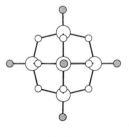

(8.3)

$[Mo_6O_{18}(NC_6H_4Me)]^{2-}$. Alternative routes are to use isocyanates (Eqn. 8.4) or amines (Eqn. 8.5).

$$[Mo_6O_{19}]^{2-} + RNCO \rightarrow [Mo_6O_{18}(NR)]^{2-} + CO_2 \quad (R = alkyl\ or\ aryl) \quad (8.4)$$

$$[Mo_6O_{19}]^{2-} + ArNH_2 \rightarrow [Mo_6O_{18}(NAr)]^{2-} + H_2O \quad (Ar = aryl) \quad (8.5)$$

By using diisocyanates or diamines, these reactions can be developed to connect together $\{Mo_6O_{18}\}$ units (e.g. Eqn. 8.6), or produce difunctionalized cages (e.g. Eqn. 8.7).

$$2[Mo_6O_{19}]^{2-} + OCN-X-NCO \rightarrow [Mo_6O_{18}(NXN)Mo_6O_{18}]^{4-} + 2CO_2 \quad (8.6)$$

X = bridging organic group, e.g. **8.5** or **8.6**

$$[Mo_6O_{18}(NXNH_2)]^{2-} + H_2NXNH_2 \rightarrow [(H_2NXN)Mo_6O_{17}(NXNH_2)]^{2-} + H_2O$$

$$X = \textbf{8.5} \quad (8.7)$$

Figures 8.13 and 8.14 show the structures of derivatives obtained using this methodology.

Comparisons of the properties (e.g. potentials for one-electron reductions) of the imido species with those of their parent polyoxometallate are consistent with the π-donor abilities of the terminal group being in the order O > NAr > NR (Ar = aryl, R = alkyl). Multifunctionalization has been achieved, with up to six (i.e. the maximum possible) terminal substituents being substituted to give the derivatives $[Mo_6O_{19-x}(NAr)_x]^{2-}$ (x = 1-6).

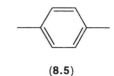

(8.4)

(8.5)

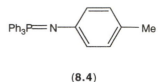

(8.6)

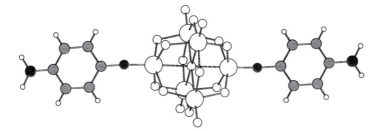

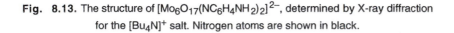

Fig. 8.13. The structure of $[Mo_6O_{17}(NC_6H_4NH_2)_2]^{2-}$, determined by X-ray diffraction for the $[Bu_4N]^+$ salt. Nitrogen atoms are shown in black.

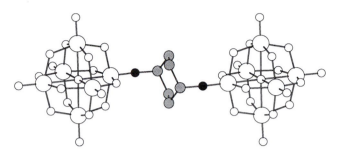

Fig. 8.14. The structure of $[Mo_6O_{18}(NC_6H_{10}N)Mo_6O_{18}]^{4-}$, determined by X-ray diffraction for the $[Bu_4N]^+$ salt; H atoms of the cyclohexane ring are not shown. Nitrogen atoms are shown in black.

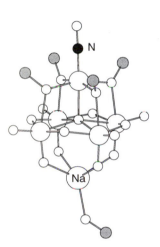

Fig. 8.15. The structure of $[Na(MeOH)][Mo_5O_{13}(\mu\text{-}OMe)_4(NO)]^{2-}$ determined by X-ray diffraction for the $[Bu_4N]^+$ salt; H atoms are omitted.

Nitrosyl derivatives have attracted significant attention, in part because the reductive nitrosylation of polyoxomolybdates by hydroxylamine has proved to be a method of preparing 'giant clusters'.[§] The reaction between hydroxylamine and $[Mo_8O_{26}]^{4-}$ in methanol leads to the formation of $[Mo_5O_{13}(\mu\text{-}OMe)_4(NO)]^{3-}$. In the salt $[Na(MeOH)][Bu_4N]_2[Mo_5O_{13}(\mu\text{-}OMe)_4(NO)]$, the solvated sodium ion interacts with the $[Mo_5O_{13}(\mu\text{-}OMe)_4(NO)]^{3-}$ cage giving a species (Fig. 8.15) which can be considered to be a complex of a lacunary-anion derived from $[Mo_6O_{19}]^{2-}$ (Fig. 8.4). In acetonitrile or dichloromethane solutions, $[Mo_5O_{13}(\mu\text{-}OMe)_4(NO)]^{3-}$ undergoes conversion into $[Mo_6O_{18}(NO)]^{3-}$; a related anion is $[W_5O_{18}Mo(NO)]^{3-}$. The introduction of the nitrosyl substituent increases the reactivity of the cage with respect to methylation by increasing the basicity of the bridging oxygen atoms adjacent to the M(NO) site. Whereas $[Mo_6O_{19}]^{2-}$ resists methylation, treatment of $[Mo_6O_{18}(NO)]^{3-}$ or $[W_5O_{18}Mo(NO)]^{3-}$ with dimethyl sulfate, $(MeO)_2SO_2$, results in the conversion of a $\mu\text{-}O$ to $\mu\text{-}OMe$ and the formation of, e.g., $[Mo_6O_{17}(\mu\text{-}OMe)(NO)]^{2-}$; a second product of the reaction is $[Mo_{12}O_{40}S]^{2-}$. In methanol solutions, the reduction of $[Mo_6O_{18}(NO)]^{3-}$ leads to the formation of $[Mo_{10}O_{25}(\mu\text{-}OMe)_6(NO)]^-$ and $[Mo_{10}O_{24}(\mu\text{-}OMe)_7(NO)]^{2-}$ which are structurally related to $[W_{10}O_{32}]^{4-}$ (Fig. 8.6).

With its 'open face', we have already illustrated (Fig. 8.15) that $[Mo_5O_{13}(\mu\text{-}OMe)_4(NO)]^{3-}$ is able to function as a tetradentate ligand. This is also observed in complexes such as $[(\eta^5\text{-}C_5Me_5)_2Rh_2(\mu\text{-}Br)\{Mo_5O_{13}(\mu\text{-}OMe)_4(NO)\}]$ (Fig. 8.16), formed in the reaction of $[Mo_5O_{13}(\mu\text{-}OMe)_4(NO)]^{3-}$ and $[\{(\eta^5\text{-}C_5Me_5)RhCl_2\}_2]$ with addition of bromide. A related complex is $[(\eta^5\text{-}C_5Me_5)Rh(H_2O)\{Mo_5O_{13}(\mu\text{-}OMe)_4(NO)\}]^-$ in which

[§] For a detailed review on giant clusters emphasizing structures and magnetic properties, see: A. Müller, F. Peters, M.T. Pope and D. Gatteschi (1998) *Chemical Reviews*, **98**, 239. For a recent example of a giant, ring-shaped polyoxomolybdate, see: A. Müller, M. Koop, H. Bögge, M. Schmidtmann and C. Beugholt (1998) *Chemical Communications*, 1501.

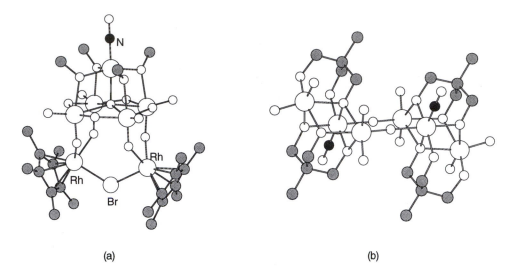

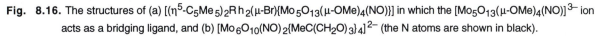

(a)   (b)

Fig. 8.16. The structures of (a) $[(\eta^5\text{-}C_5Me_5)_2Rh_2(\mu\text{-}Br)\{Mo_5O_{13}(\mu\text{-}OMe)_4(NO)\}]$ in which the $[Mo_5O_{13}(\mu\text{-}OMe)_4(NO)]^{3-}$ ion acts as a bridging ligand, and (b) $[Mo_6O_{10}(NO)_2\{MeC(CH_2O)_3\}_4]^{2-}$ (the N atoms are shown in black).

the polyoxomolydate ion acts as a didentate ligand to the Rh(II) centre. The compound $[Na(MeOH)][Bu_4N]_2[Mo_5O_{13}(\mu\text{-}OMe)_4(NO)]$ has also proved to be a useful means of accessing new coordination complexes of polyoxomolybdates, and this is illustrated by its reaction with the tripodal ligand $MeC(CH_2OH)_3$ in acetonitrile. The anion produced is $[Mo_6O_{10}(NO)_2\{MeC(CH_2O)_3\}_4]^{2-}$ and its structure (Fig. 8.16b) reveals that the square-based pyramidal Mo_5-unit of the precursor has been destroyed; the role of the four tripodal ligands in 'clamping' together the remaining metal framework is obvious from Fig. 8.16b. In a second example, the reaction of $[Na(MeOH)][Bu_4N]_2[Mo_5O_{13}(\mu\text{-}OMe)_4(NO)]$ with $[MoO_2(acac)_2]$ in acetonitrile results in the formation of $[Mo_8O_{22}(NO)_2(acac)_2]$. The structure of this product (Fig. 8.17) again confirms that significant rearrangement of the molybdenum framework has accompanied the reaction.

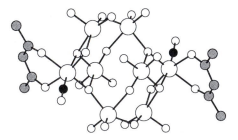

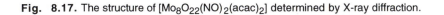

Fig. 8.17. The structure of $[Mo_8O_{22}(NO)_2(acac)_2]$ determined by X-ray diffraction.

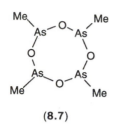

(8.7)

Detailed reviews of applications, see:
J.T. Rhule, C.L. Hill and D.A. Judd (1998) *Chemical Reviews*, **98**, 327; D.E. Katsoulis (1998) *Chemical Reviews*, **98**, 359.

Several polyoxomolybdates in which cyclopentadienyl ligands replace oxo ligands are known, and these include $[(\eta^5\text{-}C_5Me_5)Mo_6O_{18}]^-$ and $[(\eta^5\text{-}C_5Me_5)_2W_6O_{17}]$. The former complex is formed by reaction between dioxygen and $[\{(\eta^5\text{-}C_5Me_5)Mo(CO)_2\}_2]$, while oxidation of $[\{(\eta^5\text{-}C_5Me_5)W(CO)_2\}_2]$ using $Me_4As_4O_4$, **8.7**, leads to the formation of $[(\eta^5\text{-}C_5Me_5)_2W_6O_{17}]$.

8.7 Applications of polyoxometallates

Apart from their structural beauty, polyoxometallates are of immense interest because of their numerous applications arising from properties which include their stability, size (some giant clusters have *nanometre* dimensions), solubility properties, high Brønsted acidity of the conjugate acids, electron-transfer properties, and ability to model metal oxide surfaces. Although many new applications are being explored, the heteropoly blues have long been used analytically in the determination of, for example, phosphate and silicate. Present uses include those in catalysis (examples are given in Table 8.1), corrosion resistant coatings, pigments and dyes (e.g. in coatings and fibres), wood pulp bleaching, photochromic coatings in photocopiers, flame retardants, and the modification of carbon electrodes. Medical applications are of particular importance and the discovery that some polyoxometallates exhibit anti-retroviral activity (e.g. anti-HIV) is the basis for current research efforts.

Table 8.1 Examples of uses of polyoxometallates as acid catalysts.

Catalyst	Homo- or heterogeneous catalysis	Reaction
$H_3PMo_{12}O_{40}$ $H_3PW_{12}O_{40}$	Homogeneous	Propene + H_2O → propan-2-ol
$H_3PW_{12}O_{40}$	Homogeneous	Phenylacetylene + H_2O → acetophenone
$H_3PW_{12}O_{40}$	Homogeneous	Polymerization of thf
$H_3PW_{12}O_{40}$	Heterogeneous	Propan-2-ol → propene + H_2O
$H_3PMo_{12}O_{40}$	Heterogeneous	Methanol + 2-methylpropene → methyl *tert*-butyl ether
$Cs_{2.5}H_{0.5}PW_{12}O_{40}$ /Pt	Heterogeneous	Butane → 2-methylpropane

Ligand and solvent abbreviations

[acac]⁻   acetylacetonate

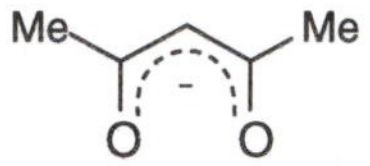

bpy   2,2'-bipyridine

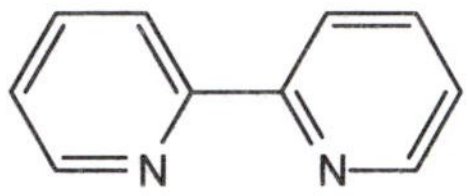

dmso   dimethyl sulfoxide

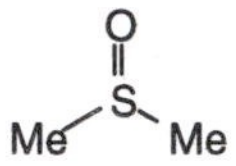

dppe   bis(diphenylphosphino)ethane

dppm   bis(diphenylphosphino)methane

en   1,2-diaminoethane (ethylenediamine)

ox²⁻   oxalate

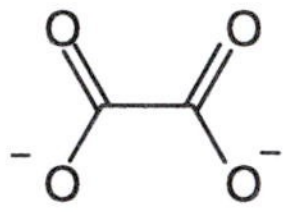

py   pyridine

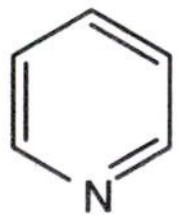

thf   tetrahydrofuran

Further reading

In addition to the specific references given in the text, the following books are recommended for further information:

R.D. Adams and F. A. Cotton, eds. (1998) *Catalysis by Di- and Polynuclear Metal Cluster Complexes*, Wiley-VCH, New York.

F. A. Cotton and R. A. Walton (1993) *Multiple Bonds between Metal Atoms*, 2nd Edn, Oxford University Press, Oxford.

F.A. Cotton and G. Wilkinson (1988) *Advanced Inorganic Chemistry*, 5th ed., Wiley, New York.

S. Cotton (1997) *Chemistry of Precious Metals*, Chapman & Hall, London.

J. Emsley (1998) *The Elements*, 3rd ed., Clarendon Press, Oxford.

J.J.R. Fraústo de Silva and R.J.P. Williams (1991) *The Biological Chemistry of the Elements*, Clarendon Press, Oxford; Chapter 17 is particularly relevant.

N.N. Greenwood and A. Earnshaw (1997) *Chemistry of the Elements*, 2nd edn, Butterworth-Heinemann, Oxford.

S.F.A. Kettle (1996) *Physical Inorganic Chemistry*, Spektrum, Oxford.

R.B. King, ed. (1994) *Encyclopedia of Inorganic Chemistry*, Wiley, Chichester (individual entries under element names).

M.T. Pope (1983) *Heteropoly and Isopoly Oxometalates*, Springer-Verlag, New York.

M.T. Pope and A. Müller, eds. (1994) *Polyoxometalates: from Platonic Solids to Anti-Retroviral Activity*, Kluwer Academic Publishers, Dordrecht.

A.G. Sharpe (1992) *Inorganic Chemistry*, 3rd ed., Longman, Harlow.

D.F. Shriver, P.W. Atkins and C.H. Langford (1994) *Inorganic Chemistry*, 2nd ed., Oxford University Press, Oxford.

Metals in action

The following articles provide accounts of aspects of some of the heavier d-block metals with an emphasis on applications or potential applications:

F. Barigelletti, L. Flamigni, J.-P. Collin and J.-P. Sauvage (1997) *Chemical Communications*, 333 – Vectorial transfer of electronic energy in rod-like ruthenium-osmium dinuclear complexes.

J. Barrett and M. Hughes (1997) *Chemistry in Britain*, **33**, number 6, 23 – A golden opportunity.

L. Davidson, Y. Quinn and D.F. Steele (1998) *Platinum Metals Review*, **42**, 90 – Ruthenium-mediated electrochemical destruction of organic wastes.

G. Denuault (1996) *Chemistry & Industry*, 678 – Microelectrodes.

R.C. Elder and K. Tepperman (1994) in *Encyclopedia of Inorganic Chemistry*, ed. R.B. King, Wiley, Chichester, vol. 4, p. 2165 – Metal-based drugs and imaging agents.

M. Green (1996) *Chemistry & Industry,* 641 –The promise of electrochromic systems.

W.P. Griffith (1992) *Chemical Society Reviews*, **21**, 179 – *Ruthenium oxo complexes as organic oxidants.*

J.N. Huiberts, R. Griessen, J.H. Rector, R.J. Wijngaarden, J.P. Dekker, D.G. de Groot and N.J. Koeman (1996) *Nature*, **380**, 213 – *Yttrium and lanthanum hydride films with switchable optical properties.*

R.W. McCabe and J. M. Kisenyi (1995) *Chemistry & Industry,* 605 – *Advances in automotive catalyst technology.*

T.R. Ralph and G.A. Hards (1998) *Chemistry & Industry,* 335 – *Fuel cells: clean energy production for the new millennium.*

T.R. Ralph and G.A. Hards (1998) *Chemistry & Industry,* 337 – *Powering the cars and homes of tomorrow*

J. Reedijk (1996), *Chemical Communications*, 801 – *Improved understanding in platinum antitumour chemistry.*

M.J. Vimy (1995) *Chemistry & Industry,* 14 – *Toxic teeth: the chronic mercury poisoning of modern man.*

M.D Ward (1996) *Chemistry & Industry,* 568 – *Current developments in molecular wires.*

C.G. Young and A.G. Wedd (1997) *Chemical Communications*, 1251 – *Molybdenum and tungsten pterin enzymes.*

World-Wide Web

The World-Wide Web is an excellent and continually changing source of information:

P. Murray-Rust, H.S. Rzepa and B.J. Whitaker (1997) *Chemical Society Reviews*, **26**, 1 – *The World-Wide Web as a chemical information tool.*

Index